Advances in Material Design for Regenerative Medicine, Drug Delivery and Targeting/Imaging

MATERIALS RESEARCH SOCIETY
SYMPOSIUM PROCEEDINGS VOLUME 1140

Advances in Material Design for Regenerative Medicine, Drug Delivery and Targeting/Imaging

Symposium held December 1–3, 2008, Boston, Massachusetts, U.S.A.

EDITORS:

V. Prasad Shastri

Albert-Ludwigs University of Freiburg
Freiburg, Germany

Andreas Lendlein

GKSS Research Centre Geesthacht GmbH
Teltow, Germany

LinShu Liu

U.S. Department of Agriculture
Wyndmoor, Pennsylvania, U.S.A.

Antonios Mikos

Rice University
Houston, Texas, U.S.A.

Samir Mitragotri

University of California - Santa Barbara
Santa Barbara, California, U.S.A.

Materials Research Society
Warrendale, Pennsylvania

Published by:

Materials Research Society
506 Keystone Drive
Warrendale, PA 15086
Telephone (724) 779-3003
Fax (724) 779-8313
Web site: http://www.mrs.org/

Manufactured in the United States of America

MATERIALS IN
REGENERATIVE MEDICINE

MACROMOLECULAR DRUG/LIPID
INTERACTION

*Invited Paper

vi

*Invited Paper

SYNTHESIS AND CHARACTERIZATION OF NANOMATERIALS FOR BIOMEDICAL APPLICATIONS

PREFACE

Many of the major breakthroughs and paradigm shifts in medicine to date have occurred due to innovations and materials and/or application/implementation of materials in clinical medicine. Artificial heart valves, implantable cardiac devices, limb prosthesis, cardiovascular stents, orthopaedic implants, and artificial skin are just a few examples of the numerous applications of materials science in modern-day medicine. The past two decades have seen the emergence of some of the most exciting avenues for innovation in clinical medicine, with materials as its core and include: sustained drug delivery systems, gene therapy, regenerative therapies, and targeting technologies primarily for imaging and tumor targeting.

Symposium HH, "Advances in Material Design for Regenerative Medicine, Drug Delivery and Targeting/Imaging," held December 1–3 at the 2008 MRS Fall Meeting in Boston, Massachusetts, covered a wide range of topics related to these various aspects of Regenerative Medicine. Session topics of this symposium included smart and active materials, controlling stem cell function, harnessing the potential of cell-based therapies in Regenerative Medicine, translation research, novel biomaterials and nano-scale systems for imaging and controlling cell function, biomaterials for drug delivery with an emphasis on intra cellular delivery and transdermal delivery, and engineered materials for gene delivery. This proceedings contains 33 peer-reviewed papers presented at the Fall Meeting, giving a representative cross-section of this great symposium.

The highlight of the symposium were the two sessions in honor of Prof. Robert Langer, Institute Professor at MIT, on the occasion of his 60th birthday, to recognize his contributions to the areas of biomaterials, drug delivery and tissue engineering. These sessions featured leaders in various highly interdisciplinary topics including two Nobel laureates and several members of national academies. Prof. Langer presented recent advances in novel biomaterials that can effectively deliver si-RNA and scaffolds that can promote spinal cord repair in primates. He also shared recent findings on the development of biomaterial niches for stem cell differentiation, a topic that was also discussed at length by Prof. Helen Blau. The role of material properties in controlling the fate of stem cells in various lineages including neurons was presented by Prof. Discher. Nobel Laureate Prof. Giaever discussed the challenges in commercializing laboratory science, and Nobel Laureate Prof. Grubbs discussed the development of new polymers for drug delivery based on ring-opening metathesis polymerization. Profs. Martin and Mooney detailed materials-based strategies to implement cell-based therapies into the clinic. The engineering of novel materials from non-natural amino acids with unique degradation behavior, with utility in corneal epithelial regeneration, was discussed by Prof. Tirrell. At the conclusion of the two special sessions and directly before his talk, Prof. Langer was awarded the inaugural Acta Biomaterialia Gold Medal to a rousing applause from the audience, for his contributions to the development of biomaterials. Other notable presentations included the development of novel polymers for imaging inflammation by Prof. Murthy, biomaterial-stem cell interactions by Prof. Hermanson, and the utility of CNT's in gene delivery by Paul Cherukuri.

We kindly acknowledge financial support for this symposium from Abbott Vascular, US Department of Agriculture, and GKSS Research Centre Geesthacht GmbH.

V. Prasad Shastri
Andreas Lendlein
LinShu Liu
Antonios Mikos
Samir Mitragotri

June 2009

MATERIALS RESEARCH SOCIETY SYMPOSIUM PROCEEDINGS

MATERIALS RESEARCH SOCIETY SYMPOSIUM PROCEEDINGS

Prior Materials Research Society Symposium Proceedings available by contacting Materials Research Society

Materials in Regenerative Medicine

Mater. Res. Soc. Symp. Proc. Vol. 1140 © 2009 Materials Research Society

Dual and Triple Shape Capability of AB Polymer Networks Based on Poly(ε-caprolactone)dimethacrylates

Marc Behl[1], Ingo Bellin[2], Steffen Kelch[3], Wolfgang Wagermaier[1] and Andreas Lendlein
[1]Institute of Polymer Research, GKSS Research Centre GmbH, Kantstr. 55, 14513 Teltow, Germany
[2]Global Polymer Research – Foams, BASF Aktiengesellschaft, GKT/F - B 001, 67056 Ludwigshafen, Germany
[3]Sika Technology AG, Tüffenwies 16, CH-8048 Zürich, Switzerland

ABSTRACT

Shape-Memory polymers are an emerging class of functionalized materials, which are able to change their change in a predefined way upon appropriate stimulation. In the past research has mostly focused on the implementation of new ways to trigger the shape-memory effect. Although the stimuli were differing, these systems were dual-shape materials. Recently, multi-phase shape-memory polymers with an additional switching phase were introduced. These materials allow the selection of the switching segment according to the requirements of the specific application. The dual shape-capability and the underlying molecular mechanisms are discussed for such multiphase polymer networks. Furthermore, these materials display triple-shape capability after appropriate programming, which enables them to change from a first shape (A) to a second shape (B) and from there to a third shape (C). Two different polymer network architectures for triple-shape polymer networks are described and investigated towards their dual and triple shape-properties. The programming of these materials can be fairly complex, therefore we will show finally an one-step programming to reach triple-shape capability.

INTRODUCTION

Shape-memory polymers have received considerable interest because of their ability to change their shape in a predefined way from a shape (A) to a shape (B) upon appropriate stimulation. This stimulation has been realized either by heat [1] or light [2]. Indirect actuation of the shape-memory effect was realized by application of irradiation with IR-light [3], of electrical fields [4], exposure to alternating magnetic fields [5, 6] or immersion in water [7]. Thus far, these systems were all dual-shape materials.

Recently, thermally-induced shape-memory polymers with an additional switching phase have been introduced [8], allowing the predefined movement by gradual temperature increase from a shape (A) to a shape (B) at a switching temperature which is determined by either the lower or the higher thermal transition. A prerequisite for this general concept is a polymer network structure with at least two phase separated domains which act as physical crosslinks. These domains have individual transition temperatures (T_{trans}), which are, depending on the type of chain segment, either a glass transition temperature (T_g) or a melting temperature (T_m). Upon cooling below T_{trans} of a specific domain, the domain solidifies and forms physical cross-links. As these additional cross-links dominate the covalent netpoints, they offer the temporary fixation of a new shape, which can be recovered into its original shape when being reheated above T_{trans}. By deformation of the polymer network and subsequent cooling in the deformed shape, this effect is used in dual-shape materials for the temporary fixation of a second shape. While shape (B) is given by the covalent crosslinks during polymer network formation, shape (A) is fixed by physical crosslinks created in a one-step thermomechanical programming process. In this programming process shape (A) is temporarily fixed by one of the two switching domains.

EXPERIMENT

Synthesis of the materials is described in references [8-10].

Cyclic, Thermomechanical Experiments. These were carried out on a Zwick (Ulm, Germany) Z005 for *CLEG*-networks and a Zwick Z1.0 for *MACL*-networks equipped with thermochambers controlled by Eurotherm control units (2216E for Z005 and 2408 for Z1.0, Eurotherm Regler, Limburg, Germany). The thermochamber for the Z005 was a T35 provided by Mytron (Heiligenstadt, Germany) while for the Z1.0 a thermochamber W91255 designed by Zwick was used. In both cases cooling was realized by a gas flow from a liquid nitrogen tank. 200 N, 100 N, and 20 N load cells were used depending on samples and temperature. Films were cut into standard samples (ISO 527-2/1BB) and strained at an elongation rate of 10 mm·min^{-1}.

In a triple shape experiment (Fig. 1) the sample is stretched at T_{high} = 70 °C for CLEG and T_{high} = 150 °C for MACL networks from ε_C, where the elongation corresponds to shape (C), to ε_B^0. Cooling with a cooling rate (β_c) of 5 K·min^{-1} to T_{low} = -10 °C for CLEG or T_{low} = 0 °C for MACL networks under stress-control results in ε_{Bload}. Afterwards the sample is held unloaded at T_{low} for 10 min. Subsequently the sample is heated to T_{mid} and after passing T_{sw} with some additional time for equilibration at T_{mid} which was in case CLEG networks 40 °C with a heating rate of β_h = 2 K·min^{-1} and in case of MACL networks 70 °C with a heating rate of β_h = 5 K·min^{-1} ε_B is reached, which is shape (B). The sample is further stretched to ε_A^0 and cooled to T_{low} under stress-control with β_c = 5 K·min^{-1}, whereas the elongation decreases to ε_{Aload}. Shape (A), corresponding to ε_A, is obtained by unloading after 10 min for *CLEG*-networks and 30 min for MACL networks. The recovery process of the sample is monitored by reheating with a heating rate of 1 K·min^{-1} from T_{low} to T_{high} while the stress is kept at 0 MPa. The sample contracts to recovered shape (B) at ε_{Brec}, which is defined as the elongation at the minimum contraction rate. Continued heating finally leads to recovery of shape (C) at ε_{Crec}. This cycle is conducted five times with the same sample.

DISCUSSION

This concept was realized in polymer network architecture, in which poly(ε-caprolactone)dimethacrlyate (PCLDMA) was copolymerized with poly(ethyleneglycol)monomethylethermethacrylate (PEGMA). These graft polymer networks are named CLEG. In such a polymer network the elasticity is mainly determined by the PCL segments which are crosslinking short polymethacrylate segments having dangling poly(ethyleneglycolmonomethylether) (PEG) segments. In CLEG networks three situations for dual shape fixation can be differentiated depending on the switching temperature [9]. I) Use of the crystallization of PCL phase for fixation of the temporary shape. II) Crystallization of the pendant PEG chain segments prevents the amorphous PCL segments from returning into the coiled state. III) Both segments can contribute to the fixation of a temporary shape. At first PCL chain segments crystallize followed by the crystallization of the PEG segments. It was shown, that every sample returns to its permanent shape, as soon as the switching temperature T_{sw} is exceeded, which is the experimentally determined temperature at the inflection point of the recovery curve. As expected, T_{sw} is close to $T_m(PEG)$ for case II and close to $T_m(PCL)$ for case I and III. As the material possesses two independent transition temperatures $T_{trans, A}$ and $T_{trans,B}$ it could also be shown that the material has triple-shape capability [8]. It should be noted that not every material presenting two independent transition temperatures displays triple shape-capability.

Triple-shape materials are materials that are able to change from a first shape (A) to a second shape (B) and from there to a third shape (C). For investigating the generality of the triple-shape capability a second multiphase polymer network was explored. In this polymer network system, called MACL, cyclohexylmethacrylate (CHMA) and PCLDMA were copolymerized. Here the polymer network structure is build by the PCLDMA and the polymerized CHMA segments. Both segments are contributing equally to the overall elasticity of the polymer network structure. In this polymer network $T_{trans,A}$ is given by the T_g of the MACL and $T_{trans,B}$ by the T_m of the PCL segments.

The triple-shape-memory effect can be characterized in specific cyclic thermomechanical experiments. In a two-step uniaxial deformation experiment the two additional shapes A and B are programmed. Afterwards, when the sample is reheated, shape B and finally shape C are recalled. These cyclic thermomechanical experiments allow the quantification of the triple-shape effect for each shape separately. While the shape fixity ratio [$R_f(X \rightarrow Y)$] describes the ability to fix shape Y after the thermomechanical programming, the shape recovery ratio [$R_r(X \rightarrow Y)$] is a measure to what extent shape Y can be recovered starting from shape X.

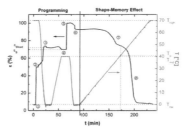

Figure 1. Cyclic, thermomechanical experiment of CL(40)EG (fifth cycle) as a function of time; solid line, strain; dashed line, temperature.

The programming of these materials can be done by different methods. Previously shown programming method I involves the stretching of the sample at T_{high} from ε_C, where the elongation corresponds to shape (C), to ε_B^0. Cooling to T_{mid} under stress-control results in ε_{Bload}. Unloading leads to ε_B, which is shape (B). The sample is further stretched to ε_A^0 and cooled to T_{low} under stress-control, whereas the elongation decreases to ε_{Aload} and shape (A), corresponding to ε_A, is obtained by unloading [8]. In contrast, a newly presented programming method II starts with stretching the sample at T_{high} from ε_C, where the elongation corresponds to shape (C), to ε_B^0. Cooling to T_{low} under stress-control results in ε_{Bload}. Afterwards the sample is held unloaded at T_{low} for 10 min. Subsequently the sample is heated to T_{mid} and after passing T_{sw} with some additional time for equilibration at T_{mid} ε_B is reached, which is shape (B). Then the sample is strained to ε_A^0. Cooling down to T_{low} leads to a decrease of the elongation because of the crystallization of the PEG segments. After unloading the sample is held at T_{low} for 10 min whereas shape (A) is obtained. Finally, after both programming methods, the sample is heated to T_{high}. A typical stress strain diagram for a CLEG polymer network is displayed in Fig. 2. The average values for the 2nd to 5th cycle of both polymer network architectures, which all exhibit the triple shape effect, are presented in table 1. The first cycle is not considered for the calculation, as this cycle is needed for the equalization of the thermomechanical history of the samples.

Table 1 Triple shape properties of polymer networks programmed with programming method II

polymer network †	ε_B^0 [%]	ε_A^0 [%]	$\overline{R}_f\,(C{\to}B)$ [%] §	$\overline{R}_f\,(B{\to}A)$ [%] §	$\overline{R}_r\,(A{\to}B)$ [%] §	$\overline{R}_r\,(A{\to}C)$ [%] §
CL(30)EG	50	100	96.8 ± 0.4	91.2 ± 0.9	82.9 ± 4.4	98.9 ± 0.8
CL(40)EG	50	100	96.6 ± 0.3	83.4 ± 1.1	77.5 ± 1.4	99.1 ± 0.6
CL(40)EG	70	100	97.2 ± 0.2	91.1 ± 1.2	90.0 ± 3.0	100.0 ± 0.9
CL(40)EG	50	120	97.5 ± 0.5	95.2 ± 1.4	84.1 ± 1.0	98.6 ± 0.8
CL(50)EG	50	100	97.5 ± 0.2	84.5 ± 2.4	73.1 ± 1.8	98.5 ± 0.1
CL(60)EG	50	100	97.6 ± 0.8	75.7 ± 1.1	-	106.0 ± 0.9
MACL(40)	50	100	96.9 ± 1.1	97.9 ± 0.4	74.7 ± 2.4	96.6 ± 4.1
MACL(45)	50	100	96.9 ± 0.6	98.7 ± 0.1	97.0 ± 1.0	97.3 ± 1.8
MACL(50)	50	100	97.0 ± 0.3	98.9 ± 0.1	101.4 ± 1.2	100.7 ± 2.5
MACL(60)	50	100	96.5 ± 0.2	99.3 ± 0.1	102.9 ± 0.7	100.0 ± 2.4

† the number in brackets indicates the content of PCL in wt%
§ average of cycles 2-5, values higher 100.0 are artefacts resulting from the tensile tester

The following equations were used to characterize the triple-shape effect.

$$R_f\,(X{\to}Y) = (\varepsilon_y - \varepsilon_x) / (\varepsilon_{yload} - \varepsilon_x) \qquad (1)$$
$$R_r\,(X{\to}Y) = (\varepsilon_x - \varepsilon_{yrec}) / (\varepsilon_x - \varepsilon_y) \qquad (2)$$

In CLEG networks $R_f(C{\to}B)$ increases with increasing PCL content as this supports the fixation of shape A, but goes along with a decrease of $R_f(B{\to}A)$. In contrast in MACL networks having triple shape capability $R_f(C{\to}B)$ does not show an influence of PCHMA content and also $R_f(B{\to}A)$ remains on a high level. This has been attributed to the difference in the network architecture, as in CLEG networks the overall elasticity is only determined by the PCL segments, while in MACL networks both segments contribute to the elasticity. All triple-shape memory materials show almost complete total recovery $R_r(A{\to}C)$. In comparison to CLEG networks the values for $R_r(A{\to}B)$ are higher in MACL networks and increase with growing PCL content. In contrast in CLEG networks $R_r(A{\to}B)$ decreases with increasing PCL content and has been explained by additional strain –induced crystallization of amorphous PCL. For MACL(50) and MACL(60) networks a complete recovery of the first temporary shape could be observed, while networks with less content of PCL show an incomplete recovery of the first shape. The comparison of the results from programming method I with the programming here presented, turns out that for CLEG and MACL with programming method II higher values of $R_f(C \rightarrow B)$ can be reached while $R_f(B \rightarrow A)$ is slightly lower or constant. In contrast, values of $R_r(A \rightarrow B)$ of CLEG networks are higher for networks with more than 40 wt% PCL with programming method II and higher for networks with less than 40 wt% PCL with programming method I. In MACL networks values for $R_r(A \rightarrow C)$ are similar with both programming methods while values for $R_r(A \rightarrow B)$ are higher for networks of more than 45 wt% PCL with programming method II.

As the programming to achieve the triple shape capability is quite complex and is even more time consuming with programming method II a one step programming method was developed [10]. This requires a polymer network architecture where both switching segments contribute to the overall elasticity as given in MACL networks. Here at $T_{high} = 150$ °C both chain segments are flexible, at $T_{mid} = 70$ °C the PCHMA chain segments are in the glassy state and upon further cooling to -10 °C the PCL chain segments form stiff flexible amorphous and rigid crystalline phases. Similar to the programming of a triple shape material for a dual shape effect [9], the samples can also be programmed in three different temperature ranges: case I $T_{low} = 70$ °C and $T_{high} = 150$ °C, case II $T_{low} = -10$ °C and $T_{high} = 70$ °C and case III $T_{low} = -10$ °C and $T_{high} = 150$ °C. The stress/strain experiments from theses three different programming procedures are depicted in figure 2 for MACL networks with 50 wt% PCL.

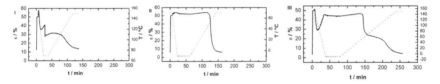

Figure 2. Strain/time diagrams (fourth cycle) obtained from cyclic, thermomechanical experiments for MACL(50): **I**: $T_{low} = 70$ °C and $T_{high} = 150$ °C; **II**: $T_{low} = -10$ °C and $T_{high} = 70$ °C; **III**: $T_{low} = -10$ °C and $T_{high} = 150$ °C.

While case I represents the dual shape capability of the PCHMA phase, case II displays the dual shape experiment for the PCL phase. When the samples were heated to T_{high} the recovery process occurs. As the time interval, which is proportional to the temperature, is significantly smaller in case II compared to case I, it can be concluded, that the melting of the PCL segments takes place over a smaller temperature interval than the softening of the PCHMA segments. In case III where both segments contribute to the fixation of the temporary shape an increase of strain, because of the thermal expansion of the sample and at 54 °C a slight decrease of the strain can be determined. This decrease has been attributed to the melting of the PCL crystallites fixing in a small amount the temporary shape. In MACL networks with 50 wt% PCL the fraction of the fixed temporary shape is considerably higher than in MACL networks with 45 wt% PCL. The recovery curves of MACL(40) for $\varepsilon_m = 30, 50, 75, 100$ and 125 % are displayed in figure 3a. While for $\varepsilon_m = 30$ % and 50 % between 0.5 and 2.0 % of the temporary shape is fixed by the PCL segments, for larger deformations ($\varepsilon_m > 75$ %) a plateau of about 3.5 % is reached (figure 3b). This has been attributed to the polymer network architecture, in which the PCL segments are acting as crosslinker and PCHMA segments form the backbone and therefore can be uncoiled to a larger extent, which can not be fixed by the PCL segments.

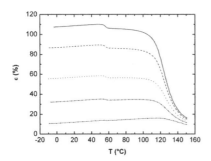 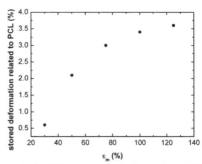

Figure 3. a Recovery curves obtained from stress-controlled, thermomechanical experiments for MACL(40) programmed with different mechanical deformations ε_m (second cycle); solid line: ε_m = 125%; dashed line: ε_m = 100%; dotted line: ε_m = 75%, dash-dotted line ε_m = 50%, dash-dot-dotted line ε_m = 30%. **b** Relationship of stored deformation of shape B to applied deformation.

CONCLUSIONS

Triple-shape materials are highly interesting materials. Proper use of programming temperature additionally enhances the versatility of these materials. Depending on the temperature used for programming either dual-shape properties with two different T_{switch} or triple-shape properties can be realized. Furthermore, by the selective choice of an appropriate programming, the triple-shape properties of such materials can be even improved, depending on the underlying molecular architecture. But this programming can be fairly complex. In certain triple-shape polymer network architecture a one-step programming was shown to be successful. A field which require complex movements of materials, e.g. for minimally invasive surgery or smart medical devices is the field of medical applications, thus seems to be very promising.

REFERENCES

1. A. Lendlein, S. Kelch, *Angew Chem, Int Ed Engl* **2002**, *41*, 2034.
2. A. Lendlein, H. Y. Jiang, O. Jünger, R. Langer, *Nature* **2005**, *434*, 879.
3. H. Koerner, G. Price, N. A. Pearce, M. Alexander, R. A. Vaia, *Nat Mater* **2004**, *3*, 115.
4. J. W. Cho, J. W. Kim, Y. C. Jung, N. S. Goo, *Macromol Rapid Commun* **2005**, *26*, 412.
5. R. Mohr, K. Kratz, T. Weigel, M. Lucka-Gabor, M. Moneke, A. Lendlein, *Proc Natl Acad Sci U S A* **2006**, *103*, 3540.
6. P. R. Buckley, G. H. McKinley, T. S. Wilson, W. Small, W. J. Benett, J. P. Bearinger, M. W. McElfresh, D. J. Maitland, *IEEE Trans Biomed Eng* **2006**, *53*, 2075.
7. B. Yang, W. M. Huang, C. Li, C. M. Lee, L. Li, *Smart Materials & Structures* **2004**, *13*, 191.
8. I. Bellin, S. Kelch, R. Langer, A. Lendlein, *PNAS* **2006**, *103*, 18043.
9. I. Bellin, S. Kelch, A. Lendlein, *J Mater Chem* **2007**, *17*, 2885.
10. M. Behl, I. Bellin, S. Kelch, A. Lendlein, *Adv Funct Mater* **2009**, *19*, 102.

Mater. Res. Soc. Symp. Proc. Vol. 1140 © 2009 Materials Research Society   1140-HH07-02

Bioreactor Systems in Regenerative Medicine: From Basic Science to Biomanufacturing

Elia Piccinini, David Wendt, Ivan Martin
Institute for Surgical Research and Hospital Management, University Hospital Basel,
Hebelstrasse 20, 4031 Basel, Switzerland

ABSTRACT

In recent years, 3D culture models have started showing their extensive potential as one of the strategic tools in the fields of tissue engineering and regenerative medicine to study various aspects of cell physiology and pathology, as well as to manufacture cell-based grafts. Given the crucial role that bioreactors play in establishing a comprehensive level of monitoring and control over specific environmental factors in 3D cultures, we review herein some of the manifold possibilities correlated with bioreactor systems in the transitional pathway between the bench and the bedside. In particular, we will draw the attention to their functions as: *1) 3D culture model systems*, enabling to outline specific aspects of the actual *in-vivo* milieu and, when properly integrated with *computational modelling* and *sensing and control* techniques, to address challenging scientific questions; *2) Graft manufacturing devices*, implementing bioprocesses so as to support safe, standardized, scaleable, traceable and possibly cost-effective production of grafts for pre-clinical and clinical use.

INTRODUCTION

Bioreactors are generally regarded as devices used in industrial biotechnological processes in order to provide a defined and controlled environment to a biological source. In Tissue Engineering (TE) applications, a "bioreactor" was initially a simple Petri dish undergoing mixing and incubated at a controlled temperature. Over time, the evolution of technical tools and the need of a better comprehension of cellular mechanisms have lead to the development of more sophisticated *in vitro* systems.

Quite recently, bioreactors have shown to be able to overcome limitations of conventional manual methods (e.g., seeding of cells into porous scaffolds and maintenance in culture of the resulting engineered constructs) to drive the development of structurally defined and functionally effective engineered grafts. In addition, bioreactors constitute technological instruments to perform controlled studies aimed at understanding the effects of biological, chemical or physical cues on cells behavior in a 3D engineered construct by maintaining specific parameters within defined ranges. These attractive features give bioreactors the capability to play an active role in facilitating the entry of TE products in the market of clinical products, or to be a new kind of product by themselves.

Defining an adequate correspondence between clinical efficacy and overall costs of TE products is one of the most strategic challenges still to address to successfully translate TE technology from bench to bedside. The difficulty of this challenge is to make a comparison of costs/ benefits ratio with traditional products/methods. In fact, manufacturing of TE products currently requires a great number of manual operations performed by trained personnel and has to be compliant with the evolving regulatory framework in terms of Quality Control (QC) and Good Manufacturing Practice (GMP) requirements.

Following this perspective, bioreactors as a means to generate and maintain a controlled culture environment and enabling directed tissue growth could represent the key element for the development of automated, standardized, traceable, cost-effective and safe manufacturing of engineered tissues for clinical applications.

In the following paragraphs, we discuss the role of bioreactors in TE applications, from basic research to streamlined tissue manufacturing processes, highlighting current scientific challenges and future clinical perspectives. In particular, we first review the key functions of bioreactors traditionally employed in research applications as 3D model systems recapitulating aspects of the actual *in vivo* milieu of specific tissues. Afterward we disclose the state of the art in computational modelling of bioreactor systems, and finally we discuss about the basic sensing techniques employed in the engineering of biological tissues to peek inside the "black box" bioreactor. Afterwards, we provide examples, potentials, and challenges of bioreactor-based manufacturing strategies for the production of tissue engineered products for clinical applications.

BIOREACTORS AS 3D *IN VITRO* MODEL SYSTEMS

Through studies of fundamental biology and tissue engineering, we have become increasingly aware that conventional 2D culture systems (i.e. Petri dishes, culture flasks) cannot adequately recapitulate the micro-environment experienced by cells *in vivo*. At the same time, it has become evident that cell differentiation and tissue development *in vivo* are strongly dependent on cell spatial organization and directional cues, and that 3D culture model systems will be vital to gain a greater understanding of basic cell and tissue physiology and pathology within the native *in vivo* micro-environment.

Urged by a rising curiosity in the mechanisms regulating cell metabolism, proliferation and differentiation, as well as cell/cell and cell/material interactions and mechano-transduction dynamics, 3D model systems have been developed to recapitulate specific aspects of the actual *in vivo* milieu of defined tissues (e.g. cartilage, bone, skeletal muscle, bone marrow). These culture systems, besides comprising the appropriate differentiated/undifferentiated cell sources and a proper 3D micro-environment (e.g., scaffolding materials), generally encompass the application of suitable *biochemical* and *physical* cues, resembling the ones sensed by the corresponding tissue *in vivo*. The comprehension of cellular mechanisms is expected to provide the key to achieve a better prediction and control on the bio-mechanical characteristics of the engineered scaffold, as recently proposed by Martin *et al.*[1]

In fact, donor variability, different cell behaviour and incomplete control over the set of physico-chemical variables contribute to make difficult, if not unmanageable, to get a defined threshold of the minimum properties (mechanical and biological) that is necessary to achieve in an engineered graft. To answer the question 'how good is good enough?', bioreactor-based *in vitro* model systems have been used to simulate the behaviour of an engineered graft upon surgical implantation and subsequent exposure to physiological loading regimes.

In one example, engineered cartilage tissues at different developmental stages were exposed to controlled loading regimes resembling a mild post-operative rehabilitation [2]. Results indicated that the response was positively correlated with the amount of glycosaminoglycans in the constructs, suggesting that a more developed engineered tissue could be better suited for early post operative loading after implantation. Reverting the concept, the same experimental setup could be exploited to identify potential regimes of physical rehabilitation which are most appropriate for grafts at a defined stage of development.

One of the current drawbacks to the implementation of bioreactor systems is the often employed trial-and-error approaches, based on expensive and inefficient attempts to optimize culture conditions, which emphasizes the need for modern rational instruments to hone the current analytical methods that are based on qualitative, endpoint, destructive, offline observations of the culture outcomes [3].

With the progressive collaboration of both biological and technical disciplines, expanding the availability of predictive tools like computational modelling should come naturally. The implementation of these instruments, aiding in sound and rational experimental design, together with the integration of technological platforms to non-invasively and non-destructively monitor the culture progression in real-time, remain critical challenges to be faced in order to allow an efficient use of bioreactor-based 3D model systems in both scientific research and clinically-compliant tissue manufacturing.

Computational models can be powerful and cost-effective tools for the design and optimization of bioreactor systems, which can be used to gain a better understanding of fundamental aspects of cell responses in complex 3D environments. Macro-scale computational models have been developed in recent years to simulate the fluid-dynamics and mass transport at the level of the bioreactor system. CFD simulations, validated through imaging techniques such as particle image velocimetry (PIV), showed that flow in spinner flasks and wavy walled bioreactors was unsteady, periodic and fully turbulent, resulting in heterogeneous fluid-induced shear stress distributions over the outer surface of the scaffolds [4;5]. Computational modeling of these systems helped to tune scaffold location and agitation rate in order to provide a more homogeneous shear stress distribution over the constructs [4]. Similar approaches, based on macro-scale models, have been carried out to evaluate the effect of the overall scaffold shape (spheroid versus cylinder) on the wall shear stress distributions within rotating wall vessel bioreactors [6] and to simulate oxygen transport and velocity/shear profiles over the surfaces of constructs placed at various locations within a concentric cylinder bioreactor [7].

In an attempt to better characterize the local hydrodynamic environment experienced by the cells (i.e., within the scaffold *pores*), micro-scale CFD models have been developed based on idealized scaffold structures with well defined simple pore architectures. Starting first with a 2D micro-scale model (a mesh scaffold subject to direct perfusion) [8] , a 3D model of a polymeric foam scaffold was later developed, even if the complex and random pore size and distribution had to be simplified as a honeycomb-like pattern [9]. The simulations showed that foam pore sizes strongly influenced the predicted average shear stress, whereas the porosity strongly affected the statistical distribution of the shear stresses but not the average value. These results served as a powerful tool for defining scaffold design criteria [10].

To overcome the limitations due to the simplification of scaffolds, micro-computed tomography (μCT) reconstructions of 3D of scaffolds can provide more realistic structures of the porous microarchitecture. CFD simulations have been used to predict local velocity and shear profiles throughout the actual pore microarchitecture of perfused scaffolds [11-15]. Interestingly, simulations of flow through a foam scaffold having high pore interconnectivities, which were based on either a μCT model [11] or a simplified geometry model [10], predicted similar globally averaged shear stresses. However, simulations based on the μCT-based model could reveal a more variable local shear distribution within individual pores and among different pores throughout the foam. On the other hand, simulations predicted a wide range of shear stresses, pressures, and fluid velocities within a calcium phosphate cement scaffold, which had many non-interconnected pores, highlighting the potential importance of pore interconnectivity [15]. μCT-based models have been used also to predict the local oxygen profiles within a cell-seeded construct during the initial stage of perfusion culture [12].

A significant limitation of these micro-scale computational approaches, based on a pre-defined scaffold geometry and cell population, is that the models are generally relevant only at the initial stage of the culture. At later time points, as cells proliferate/migrate and extracellular matrix is deposited, CFD models would need to account for the associated changes in the scaffold's effective pore microstructure and the resulting modified flow profiles. Moreover, models of mass transport would not only have to consider the altered flow patterns, but may also need to consider changes in nutrient consumption, waste production and possible changes in cell metabolism due to cell proliferation and differentiation. Some of the attempts carried out to develop models able to include cell/tissue dynamics include: modeled cell growth kinetics as a function of local cell number and nutrient concentration [16-18], GAG distribution based on local oxygen concentrations [19;20], temporal development of oxygen and cell density profiles within engineered cartilage constructs [21-23], cell automata models in which cell movements and cell proliferation are taken into account [24-28], and a mechano-regulation model to predict best mechanical properties of a scaffold to promote recruitment, proliferation and differentiation of mesenchymal stem cells from the bone marrow within osteochondral defects [29].

Together with *in silico* models, useful to rationalize the experimental approach in the field of TE but until now far under utilized in TE., another crucial aspect that still needs further development is the sensing of culture parameters like O_2, CO_2, pH, nutrients and products concentrations. Acquisition of data for the whole duration of the culture, rather than end-point analysis, can provide an indispensable means to clarify still unknown aspects of the cellular response in dynamic culture conditions, as well as playing an integral role in the automation and in-process control of tissue manufacturing processes.

Monitoring of 3D cultures has generally been hindered by the lack of adequate commercially available sensors, in particular regarding dimensions (sensors are usually too big), flow-rate independent response, lifetime long enough to sustain a complete culture period, low-cost and sterility. Invasive sensors (placed directly inside the culture chamber), achieve high precision and accuracy but are also usually costly for a single use (often based on optical and electrochemical principles); non-invasive sensors are useful if the bioreactor walls are transparent to the investigating wave (usually based on ultrasounds or optical methods like fluorimetry, spectrophotometry, Doppler Optical Coherence Tomography (DOCT) [30]); "offline analysis" give advantages in terms of less limitations for the kind of sensors used but disadvantages mainly due to the delay introduced by sampling operations.

While methods to monitor parameters directly associated with the construct itself is still at the earliest phase of development, the opportunity to non-invasively and non-destructively monitor the construct in real time would have enormous benefits in terms of costs, time, and quality control. Attempting to get an insight on cell function within the engineered construct itself, some groups have started to make efforts towards this direction [31-33]. Regarding the monitoring of morphological properties of engineered constructs, successful results have been obtained using Optical Coherence Tomography to follow the production of extracellular matrix [34], as well as using μCT imaging systems to evaluate mineralization within a cell-seeded rapid prototyped scaffold [35].

12

The establishment of well-defined and controlled bioreactor-based 3D culture model systems, supported by modeling efforts and sensing technologies, will be key to gain deep insight into the mechanisms of tissue development at the research level. Consequently, these systems may provide advanced technological platforms for the achievement of more applicative, high-throughput objectives, e.g. enabling drug screening and toxicology studies, fostering the development of new, rational design criteria for advanced biomaterials/implants, as well as allowing functional quality control of engineered tissues. Besides limiting the recourse to complex, costly and ethically questionable *in vivo* experiments in animal models, such an approach would establish the basis for safe and standardized manufacture of grafts for clinical applications.

BIOREACTORS AS GRAFT MANUFACTURING DEVICES

Despite the many progresses obtained during the last two decades in the field of cell-based therapeutic approaches, traditional products and surgical methods have still not been revolutionized (neither flanked) by TE products. Some of the main reasons are related to the long validation time before the introduction into the market and the need to show an improvement of any new product compared to pre-existent therapies in terms of repair efficiency, pain/morbidity reduction, and overall costs. In particular, TE products should demonstrate the capability of being implemented in industrial like processes, meaning reduction of costs through the substitution of manual operations with automated, standardized, cost-effective procedures.

Citing a couple of examples, Carticel® by Genzyme Tissue Repair (Cambridge, MA, USA) and Hyalograft C^{TM} Fidia Advanced Biopolimers (Abano Terme, Italy) are TE product based on cultivation of cells and preparation of the grafts for the following implant in centralized cell culture facilities. However, routine culture operations are still now mainly performed manually, with potential drawbacks such as operator variability, cross-contamination risks, limited scale-up possibilities, and high costs. As an alternative, automated and closed bioreactor systems, which minimize operator dependence, may provide advantageous in terms of safety, costs, GMP regulations and scale-up possibilities.

Following this approach, Hamilton AG (Bonaduz, Switzerland), and The Automation Partnership, (Royston, UK), represent a couple of examples of integration between cell culture processes and automation through the employment of robots for some laborious operations. But still, costs-benefits parameters would need to be defined in relation to the elevated entry investment, and moreover, automation of 2D cell cultures represents only one component of the numerous key bioprocesses required to generate three-dimensional tissue grafts. One of the first examples toward an automated system able to culture cells in a 3D environment was implemented by Advanced Tissue Sciences for the production of Dermagraft® [36;37], an allogenic product manufactured with dermal fibroblasts. Although the system was affected by some drawbacks, it gave the possibility to easily increase the number of cultured grafts in a scale-up approach to increase production volumes, thus representing a concrete attempt to translate TE know-how in a functional and profitable high-tech product. In a different perspective, based on the "scaling-out" of a manufacturing system [38], the concept of ACTES (Autologous Clinical Tissue Engineering System), previously under development Millennium Biologix, was a modular system with several parallel and independent bioreactors for autologous cell sources. ACTES would allow cultivation of cells from different donors in compartmentalized modules able to perform the complete processing of cells, starting with the harvesting of cells from the biopsy to the cultivation in a 3D environment.

In parallel to automation, an approach more related to industries and TE firms, basic research is trying to figure out streamlined processes in order to simplify current procedures or to reduce time and costs necessary to manufacture an engineered graft. A bioreactor-based concept was recently described by Braccini *et al.* for the engineering of osteoinductive bone grafts in a streamlined process, bypassing the conventional phase of selection and cell expansion on plastic dishes [39]; a concept particularly appealing to implement in a simple and streamlined manufacturing process. A similar strategy, aimed to simplify the overall process, was suggested by Scherberich *et al.*, in which adipose tissue-derived cells directly cultured without monolayer expansion are able to generate both vasculogenic and osteogenic constructs, and obtaining a reduction of culture time from 3 weeks to 5 days due to a higher frequency of progenitor cells [40] occurring in adipose tissue with respect to bone marrow.

Increasing emphasis has been brought on reducing operational times and manipulations of biological materials through the employment of decentralized facilities in alternative to centralized, specialized TE firms delegated to manipulate cells, aiming at the ultimate goal of an intra-operative therapy in which a graft able to provide minimal requirements is cultivated directly *in vivo* in the final implant place. The role of bioreactors in this application could consist in automating the separation of the cells from a biopsy and the seeding on a scaffold for the successive implant. Thus, shortening and simplification of procedures represent key factors for a safer and wider implementation of TE products in the clinics.

CONCLUSIONS

The ex vivo generation of living tissue grafts has presented new biological and engineering challenges for establishing and maintaining cells in three-dimensional cultures, therefore necessitating the development of new biological models as compared to those long established for traditional cell culture. In this context, bioreactors represent a key tool in the tissue engineering field, from the initial phases of basic research through the final manufacturing of a product for clinical applications.

Changing the trial-and-error approach with a more rational strategy by means of predictive tools based on computational models will sensibly reduce costs and times related to the research, while the development of sensing systems will open a window into the "black box" constituted by the bioreactor. We are still in a phase in which the strategic parameters that lead to cell growth, proliferation and differentiation are not well defined. Bioreactor systems represent an exceptional tool to define these parameters, in order to use them in the future to redesign bioreactor themselves in a simple and effective concept.

Collaborations between academic institutions, providing fundamental aspects of basic research and fundamental requisites for bioreactor systems, and industrial partners, helping with automation and the creation of simple user interfaces, would be of great benefit for developing new bioreactor systems and their use to finalize a couple of decades of research in the TE field toward the manufacturing of cell based products in a competitive way.

REFERENCES

1. I. Martin, D. Wendt, and M. Heberer, Trends Biotechnol. **22** (2), 80-86 (2004).
2. O. Demarteau, D. Wendt, A. Braccini, M. Jakob, D. Schafer, M. Heberer, and I. Martin, Biochem. Biophys. Res. Commun. **310** (2), 580-588 (2003).
3. C. Candrian, D. Vonwil, A. Barbero, E. Bonacina, S. Miot, J. Farhadi, D. Wirz, S. Dickinson, A. Hollander, M. Jakob, Z. Li, M. Alini, M. Heberer, and I. Martin, Arthritis Rheum. **58** (1), 197-208 (2007).
4. B. Bilgen and G. A. Barabino, Biotechnol. Bioeng. **98** (1), 282-294 (2007).
5. P. Sucosky, D. F. Osorio, J. B. Brown, and G. P. Neitzel, Biotechnol. Bioeng. **85** (1), 34-46 (2004).
6. R. A. Gutierrez and E. T. Crumpler, Ann. Biomed. Eng **36** (1), 77-85 (2008).
7. K. A. Williams, S. Saini, and T. M. Wick, Biotechnol. Prog. **18** (5), 951-963 (2002).
8. M. T. Raimondi, F. Boschetti, L. Falcone, G. B. Fiore, A. Remuzzi, E. Marinoni, M. Marazzi, and R. Pietrabissa, Biomech. Model. Mechanobiol. **1** (1), 69-82 (2002).
9. M. T. Raimondi, F. Boschetti, L. Falcone, F. Migliavacca, A. Remuzzi, and G. Dubini, Biorheology **41** (3-4), 401-410 (2004).
10. F. Boschetti, M. T. Raimondi, F. Migliavacca, and G. Dubini, J. Biomech. **39** (3), 418-425 (2006).
11. M. Cioffi, F. Boschetti, M. T. Raimondi, and G. Dubini, Biotechnol. Bioeng. **93** (3), 500-510 (2006).
12. M. Cioffi, J. Kueffer, S. Stroebel, G. Dubini, I. Martin, and D. Wendt, J. Biomech. **(In Press)** (2008).
13. B. Porter, R. Zauel, H. Stockman, R. Guldberg, and D. Fyhrie, J. Biomech. **38** (3), 543-549 (2005).
14. M. T. Raimondi, M. Moretti, M. Cioffi, C. Giordano, F. Boschetti, K. Lagana, and R. Pietrabissa, Biorheology **43** (3-4), 215-222 (2006).
15. C. Sandino, J. A. Planell, and D. Lacroix, J. Biomech. **41** (5), 1005-1014 (2008).
16. C. J. Galban and B. R. Locke, Biotechnol. Bioeng. **56** (4), 422-432 (1997).
17. C. J. Galban and B. R. Locke, Biotechnol. Bioeng. **65** (2), 121-132 (1999).
18. C. J. Galban and B. R. Locke, Biotechnol. Bioeng. **64** (6), 633-643 (1999).
19. B. Obradovic, J. H. Meldon, L. E. Freed, and G. Vunjak-Novakovic, AICHE J. **46** (9), 1860-1871 (2000).
20. I. Martin, B. Obradovic, L. E. Freed, and G. Vunjak-Novakovic, Ann. Biomed. Eng **27** (5), 656-662 (1999).
21. M. C. Lewis, B. D. Macarthur, J. Malda, G. Pettet, and C. P. Please, Biotechnol. Bioeng. **91** (5), 607-615 (2005).
22. J. Malda, D. E. Martens, J. Tramper, C. A. van Blitterswijk, and J. Riesle, Crit Rev. Biotechnol. **23** (3), 175-194 (2003).
23. J. Malda, T. B. Woodfield, d. van, V, F. K. Kooy, D. E. Martens, J. Tramper, C. A. van Blitterswijk, and J. Riesle, Biomaterials **25** (26), 5773-5780 (2004).
24. Y. Lee, S. Kouvroukoglou, L. V. McIntire, and K. Zygourakis, Biophys. J. **69** (4), 1284-1298 (1995).
25. C. A. Chung, C. W. Yang, and C. W. Chen, Biotechnol. Bioeng. **94** (6), 1138-1146 (2006).

26. L. E. Freed, G. Vunjak-Novakovic, J. C. Marquis, and R. Langer, Biotechnol. Bioeng. **43** (7), 597-604 (1994).
27. G. Cheng, B. B. Youssef, P. Markenscoff, and K. Zygourakis, Biophys. J. **90** (3), 713-724 (2006).
28. F. Galbusera, M. Cioffi, M. T. Raimondi, and R. Pietrabissa, Comput. Methods Biomech. Biomed. Engin. **10** (4), 279-287 (2007).
29. D. J. Kelly and P. J. Prendergast, Tissue Eng **12** (9), 2509-2519 (2006).
30. C. Mason, J. F. Markusen, M. A. Town, P. Dunnill, and R. K. Wang, Biosens. Bioelectron. **20** (3), 414-423 (2004).
31. J. S. Stephens, J. A. Cooper, F. R. Phelan, Jr., and J. P. Dunkers, Biotechnol. Bioeng. **97** (4), 952-961 (2007).
32. G. P. Botta, P. Manley, S. Miller, and P. I. Lelkes, Nat. Protoc. **1** (4), 2116-2127 (2006).
33. O. A. Boubriak, J. P. Urban, and Z. Cui, J. R. Soc. Interface **3** (10), 637-648 (2006).
34. P. O. Bagnaninchi, Y. Yang, N. Zghoul, N. Maffulli, R. K. Wang, and A. J. Haj, Tissue Eng **13** (2), 323-331 (2007).
35. B. D. Porter, A. S. Lin, A. Peister, D. Hutmacher, and R. E. Guldberg, Biomaterials **28** (15), 2525-2533 (2007).
36. W. A. Marston, J. Hanft, P. Norwood, and R. Pollak, Diabetes Care **26** (6), 1701-1705 (2003).
37. G. K. Naughton, Ann. N. Y. Acad. Sci. **961**, 372-385 (2002).
38. C. Mason and M. Hoare, Regen. Med. **1** (5), 615-623 (2006).
39. A. Braccini, D. Wendt, C. Jaquiery, M. Jakob, M. Heberer, L. Kenins, A. Wodnar-Filipowicz, R. Quarto, and I. Martin, Stem Cells **23** (8), 1066-1072 (2005).
40. A. Scherberich, R. Galli, C. Jaquiery, J. Farhadi, and I. Martin, Stem Cells **25** (7), 1823-1829 (2007).

Mater. Res. Soc. Symp. Proc. Vol. 1140 © 2009 Materials Research Society   1140-HH03-01

Shape-memory properties of multiblock copolymers consisting of poly(ω-pentadecalactone) hard segments and crystallizable poly(ε-caprolactone) switching segments

Karl Kratz[1,2], Ulrike Voigt[1], Wolfgang Wagermaier[1], and Andreas Lendlein[1,2]

[1]Research Centre Geesthacht GmbH, Inst. for Polymer Research, Kantstr.55, 14513 Teltow, Germany
[2]Berlin-Brandenburg Centre for Regenerative Therapies, Augustenburger Platz 1, 13353 Berlin, Germany

ABSTRACT

A series of resorbable multiblock copolymers (PDLCL) containing poly(ω-pentadecalactone) hard segments (PPDL) and crystallizable poly(ε-caprolactone) switching segments (PCL) showing a thermally-induced shape-memory effect were synthesized via co-condensation of the oligomeric macrodiols with an aliphatic diisocyanate. Thermal and mechanical properties at different temperatures were explored for copolymers with different hard segment content. All PDLCL polymers exhibit excellent shape-memory properties with shape recovery rates R_r in the range of 91% to 98% determined in the 2nd cycle under stress-free conditions. The switching temperature T_{sw} and the temperature $T_{\sigma,max}$ related to the stress maximum when heated under constant strain conditions as well as the temperature interval ΔT_{rec} spanned during shape recovery was found to be independent from copolymer composition.

INTRODUCTION

Shape-memory polymers have attracted tremendous interest because of their scientific and technological significance [1, 2]. In particular multifunctional resorbable multiblock copolymers with a unique linear mass loss are reported as promising candidates for medical applications e.g. an intelligent surgical suture [3-6]. In order to tailor shape-memory polymers to the specific requirements of a potential application a detailed knowledge about the correlation between the shape-memory properties and the thermomechanical programming process for creating the dual shape capability (e.g. fixation temperature T_{low}) is required. Characteristics of shape-memory effect (SME) are the switching temperature T_{sw} that needs to be exceeded to recall a memorized shape under stress-free recovery conditions and the temperature $T_{\sigma,max}$ related to the stress maximum when heated under constant strain conditions. As the shape recovery spans a certain temperature range ΔT_{rec}, T_{sw} is determined as the inflection point in the strain-temperature recovery curve.

In this work we present a series of PDLCL multiblock copolymers containing PPDL hard segments and crystallizable PCL switching segments showing a thermally-induced shape-memory effect. We chose the co-condensation method according to ref. [3] to prepare PDLCL multiblock copolymers by coupling the telechelic macrodiols with 2,2(4),4-trimethylhexanediisocyanate (TMDI) [7]. PPDL-diol as telechelic precursor with M_n *(app)* = 5600 g·mol^{-1} and T_m = 91 °C was prepared by ring-opening polymerization of ω-pentadecalactone with ethylene glycol as initiator [8]. A crystallizable switching segment was realized by incorporation PCL-diol precursor with M_n = 3000 g·mol^{-1} and T_m = 43 °C.

The mechanical properties of PDLCL copolymers were analyzed under variation of the hard to switching segment ratio. Shape-memory properties of PDLCL polymers were investigated by cyclic, thermomechanical tensile tests carried out on a tensile tester equipped with a thermo chamber. Each cycle consists of a thermomechanical programming process (*TPP*), where the temporary shape is created, and a recovering module where the recovery of the permanent shape is induced. Two different recovery modules were applied: stress-free ($\sigma = 0$ MPa) and constant strain conditions.

EXPERIMENTAL DETAILS

The synthesis of multiblock copolymers is exemplarily described for PDLCL040: 20 g PPDL-diol, 30 g PCL-diol (Solvay, Warringthon, UK) and 4,4 g TMDI (> 99%, Fluka, Taufkirchen, Germany) were reacted in 1,2-dichlorethane at 80 °C for 10 days. The product was obtained by precipitation in methanol at -10 °C (yield 91%).

Test specimens for DMTA, tensile testing and WAXS experiments were produced with a polymer press (type 200 E, Dr. Collin, Ebersberg, Germany) at 100 °C under a pressure of 90 bar for four minutes. The film thickness was typically between 0.4 and 0.6 mm. Specimen type 1BB according to European Standard for tensile test (DIN EN ISO1BB) were cut with a punching tool, $I_0 = 20$ mm, width 2 mm.

DMTA measurements were performed on a Gabo (Ahlden, Germany) Eplexor® 25 N. All experiments were performed in temperature sweep mode with a constant heating rate of 2 K·min^{-1} and an oscillation frequency of 10 Hz in the temperature range from -100 °C to 170 °C. $T_{max,\delta}$ is the maximum peak temperature of the *tan δ* curve.

All mechanical and cyclic, thermomechanical tensile tests were performed using tensile tester Zwick Z1.0 or Z005 (Zwick, Ulm, Germany) equipped with a thermo chamber and temperature controller (Eurotherm Regler, Limburg, Germany). Cyclic thermomechanical tests are repeated three times per experiment. The first cycle was applied as pre-conditioning procedure, while the data were determined in the 2nd cycle.

The WAXS measurements were carried out with an X-ray diffraction system Bruker D8 Discover with a two-dimensional detector from Bruker AXS (Karlsruhe, Germany). The X-ray generator, producing copper K-alpha radiation with a wavelength of 0.154 nm, was operated at a voltage of 40 kV and a current of 40 mA. A graphite monochromator and a pinhole collimator with an opening of 0.8 mm defined the optical and geometrical properties of the beam. Samples were illuminated for 2 minutes in transmission geometry and the diffraction images were collected at a sample-to-detector distance of 15 cm. The two-dimensional diffraction images were integrated to obtain plots with intensity versus diffraction angle (2-theta). These profiles were analyzed using the Bruker software TOPAS to determine the degree of crystallinity, which is the ratio between areas of crystalline peaks to the total area below the diffraction curve.

RESULTS AND DISCUSSION

The apparent number average molecular weight M_n *(app)* and the polydispersity index (PD) of the different PDLCLs were determined by gel permeation chromatography (GPC). The obtained values for M_n *(app)* are between 58,000 g·mol^{-1} and 116,000 g·mol^{-1} with PD in the range form 2.0 to 3.3. The PPDL/PCL ratio was confirmed by ^{1}H–NMR from the integration ratio of the PPDL signal at 1.25 ppm and PCL signal at 1.39 ppm. The thermal properties were investigated by DSC and dynamic mechanical analysis at varied temperature (DTMA). In all DSC thermograms two distinct melting transitions are observed for the copolymers. The melting temperature $T_{m,PCL}$ at around 44 °C belongs to crystalline PCL phase and $T_{m,PPDL}$ around 82 °C corresponds to the crystallites of the PPDL phase. By increasing of the PPDL hard segment content in the multiblock copolymer from 30 wt% to 60 wt%, $T_{max\delta}$ attributed to the glass transition temperature, is shifted from 33 °C to 22 °C. That might be an indication for the formation of a mixed amorphous phase. Wide angle x-ray scattering (WAXS) enabled determination of the overall degree of crystallinity (DOC) of PCL and PPDL having an isomorphous crystal structure [9,10]. Samples exhibit uniform DOC values of about 33% to 39% before thermomechanical treatment.

Table 1. Sample composition, molecular weight and thermal properties of PDLCL multiblock copolymers determined by GPC, DSC and DMTA.

Sample ID[a]	PPDL [w%]	$M_n (app)$[b] [g·mol^{-1}]	PD[b]	DOC[c] [%]	$T_{max\delta}$[d] [°C]	$T_{m,PCL}$[e] [°C]	$T_{m,PPDL}$[e] [°C]
PDLCL060	60	86000	3.3	37	-22	42	82
PDLCL050	50	59000	2.0	33	-25	44	86
PDLCL040	40	58000	2.2	39	-33	43	82
PDLCL030	30	116000	2.7	37	-33	43	81

[a] The three-digit number gives the weight content of PPDL in wt%. [b] $M_n (app)$ (molecular weight) and $PD = M_w/M_n$ (polydispersity index) determined by GPC measurements. [c] Degree of crystallinity determined in WAXS experiments [d] Peak temperature of $tan\delta$ observed by DMTA attributed to glass transition. [e] Melting temperatures of PPDL and PCL domains, respectively, analyzed by DSC.

Mechanical properties

Mechanical properties were determined by tensile tests at room temperature (25 °C), at the deformation temperature to $T^{deform} = 55°C$ and recovery temperature $T_{high} = 65$ °C according to the temperatures applied in cyclic thermomechnical tests and the results are summarized in Table 2.

Table 2. Mechanical properties of PDLCL copolymers determined by tensile test at different temperatures.

	$T = 25$ °C		$T = 55$ °C		$T = 65$ °C	
Sample ID[a]	E[b] [MPa]	ε_R[b] [%]	E[c] [MPa]	ε_R[c] [%]	E[d] [MPa]	ε_R[d] [%]
PDLCL060	64 ± 9	1000 ± 200	19 ± 3	810 ± 110	6.1 ± 0.5	690 ± 90
PDLCL050	70 ± 8	1100 ± 200	15 ± 3	900 ± 200	4.1 ± 0.8	460 ± 140
PDLCL040	55 ±10	950 ± 60	10 ± 2	700 ± 200	4.3 ± 0.8	470 ± 80
PDLCL030	60 ± 10	1080 ± 90	7 ± 1	1090 ± 110	4.3 ± 0.3	680 ± 60

[a] The three-digit number gives the weight content of PPDL in wt%. [b] Young´s modulus and elongation at break ε_R determined by tensile testing at room temperature. [c] Young´s modulus and elongation at break ε_R determined by tensile testing at $T^{deform} = 55°C$. [c] Young´s modulus and elongation at break ε_R determined by tensile testing at $T_{high} = 65°C$.

At 25 °C all PDLCL copolymers exhibit similar mechanical properties independent from their composition with values for the Young´s modulus in the range of 55 MPa to 70 MPa and an elongation at break ε_R of approximately 1000% determined in tensile tests. At 55 °C and 65 °C the Young´s modulus decreased significantly compared to the values at 25 °C caused by melting of the PCL switching domains and a reduction of the overall crystallinity. At $T^{\text{deform}} = 55°C$ the Young´s modulus increases systematically with increasing hard segment content from 7 MPa for PDLCL030 to 19 MPa for PDLCL060. The same trend was observed for $T_{\text{high}} = 65$ °C.

At 55 °C all copolymers show a decrease in ε_R with values in the range of 700% to 1090% and respectively 460% and 690% at 65 °C. This can again be attributed to the already molten PCL domains.

Shape-memory properties

Shape-memory properties of PDLCL were investigated by cyclic, thermomechanical tensile tests carried out on a tensile tester equipped with a thermo chamber. Each cycle consists of a four step thermomechanical programming process (*TPP*) where the temporary shape is created followed by a recovering module where the recovery of the permanent shape is induced. *TPP* starts with heating the specimen from room temperature to the recovery temperature $T_{\text{high}} =$ 70 °C (step 1), then the sample is cooled to $T^{\text{deform}} = 55°C$ (step 2), where deformation to $\varepsilon_m =$ 150% is performed and the strain is kept constant for 5 minutes to allow relaxation (step 3). In the following the sample is cooled down to $T_{\text{low}} = 0$ °C for fixation of the temporary shape and equilibrated for 10 min at T_{low}. Finally the stress is removed (step 4). The elongation of the sample in the temporary state is given by ε_u. After finishing *TPP* the fixed temporary shape is characterized by the shape fixity rate R_f defined as the ratio of the elongation in the tension-free state after step 4 (ε_u) and the extension applied in step 3 (ε_m) [1].

For restoration of the original shape two different recovery modules were explored, the recovery under stress-free conditions ($\sigma = 0$ MPa) quantified by T_{sw} and shape fixity rate R_f, which is a measure for the ability of the material to regain its original shape represented by the tension-free strain value after the completion of the recovery process is ε_p. Both values are obtained from the strain-temperature recovery curve [9]. Under constant strain recovery conditions, where the strain level is kept constant, so that the length of the sample can not change, the maximum stress σ_{max} generated during the shape memory effect as well as the corresponding temperature ($T_{\sigma,\text{max}}$) are determined from the stress-temperature recovery curve [10].

All PDLCL polymers exhibit excellent shape-memory properties with shape recovery rates R_r in the range of 91% to 98% and values for the shape fixity rate R_f from 89% to 99% determined in the 2nd cycle. Both values for R_f as well for R_r were found to decrease with decreasing amount of PCL in the copolymer responsible for the formation of switching domains as expected (see Fig. 1). The obtained switching temperature under stress-free recovery as well as the temperature related to the stress maximum under constant strain conditions are independent from copolymer composition at values $T_{\text{sw}} = 53$ °C and $T_{\sigma,\text{max}}$ between 62 °C to 65 °C, while ΔT_{rec} spans a temperature interval of almost 25 K.

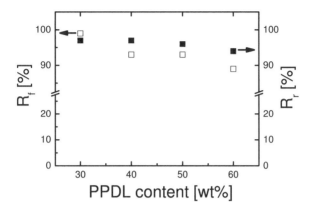

Figure 1. Shape recovery rate R_r and shape fixity rate R_f determined in cyclic thermomechanical experiments (2nd cycle) for PDLCL copolymers with different composition

CONCLUSIONS

A series of PDLCL multiblock copolymers containing poly(ε-caprolactone) switching segments and poly(ω-pentadecalactone) hard segments with various compositions were prepared. Thermal and mechanical properties at different temperatures were investigated with respect to the different hard segment content. These copolymers display excellent shape-memory properties determined under stress-free recovery conditions as well as under constant strain conditions. While T_{sw} and $T_{\sigma,max}$ were almost independent from the copolymer composition the values for R_f and R_r decreased slightly with decreasing amount of PCL.

ACKNOWLEDGMENTS

The authors thank H. Kosmella and M. Zierke (Institute of Polymer Research) for assistance with thermomechanical tests and synthesis.
Authors are grateful to the German Federal Ministry of Education and Research for providing support within the BioFuture Award No. 0311867.

REFERENCES

1. A. Lendlein, S. Kelch, *Angew. Chem. Int. Edit.* **41**, 2034 (2002).
2. M. Behl, A. Lendlein, *Soft Matter* **3**, 58 (2007).
3. A. Lendlein, R. Langer, *Science* **296**, 1673 (2002).

4. R. Mohr, K. Kratz, T. Weigel, M. Lucka-Gabor, M. Moneke, A. Lendlein, *Proc. Natl. Acad. Sci. USA* **103**(10), 3540-3545 (2006).
5. A. Kulkarni, J. Reiche, J. Hartmann, K. Kratz, A. Lendlein, *Eur. J. Pharm. Biopharm.* **68**(1),46-56 (2008).
6. J. Reiche, A. Kulkarni, K. Kratz, A. Lendlein, *Thin Solid Films* **516**, 8821-8828 (2008).
7. A. Lendlein, A. Schmidt, K. Kratz, J. Schulte, US 2004/0014929.
8. A. Lendlein, P. Neuenschwander, U. W. Suter, *Macromol. Chem. Phys.* **201**, 1067 (2000).
9. H. Bittiger, R. H. Marchessault, W. D. Niegisch, *Acta Crystallogr. B* **26**, 1923 (1970).
10. M. Gazzano, V. Malta, M. L. Focarete, M. Scandola, R. A. Gross, *J. Polym. Sci. Pol. Phys.* **41**, 1009 (2003).
11. S. Kelch, S. Steuer, A. M. Schmidt, A. Lendlein, *Biomacromolecules* **8**, 1018 (2007).
12. K. Gall, C. M. Yakacki, Y. P. Liu, R. Shandas, N. Willett, K. S. Anseth, *J. Biomed. Mater. Res. A* **73A**, 339 (2005).
13. B. Bogdanov, V. Toncheva, E. Schacht, *Macromol. Symp.* **152**, 117 (2000).

Mater. Res. Soc. Symp. Proc. Vol. 1140 © 2009 Materials Research Society   1140-HH06-08-DD03-08

Different Macrophage Responses on Hydrophilic and Hydrophobic Carbon Nanotubes

Young Wook Chun, Dongwoo Khang and Thomas J. Webster
Division of Engineering, Brown University, 184 Hope Street
Providence, RI 02912 U.S.A.

ABSTRACT

Purified carbon nanotubes (with removed toxic catalytic particles) have been considered as novel materials for drug delivery and for generating artificial organs more efficiently due to their unique surface features. Traditionally, the surface chemistry of carbon nanotubes has been modified through various functionalization strategies to increase biocompatibility. Importantly, modulating the intrinsic material surface energy of carbon nanotubes (without functionalization, thus, establishing permanent, non degradable chemical, and physical surface properties) can potentially reduce an immune response mediated by macrophages. Herein, we report macrophage responses on different surface energy carbon nanotubes while keeping their nanoscale surface roughness. Specifically, interactions of ultra hydrophobic (bare or unmodified) and hydrophilic carbon nanotubes (due to the formation of oxide layers) with macrophages were investigated. All results clearly support that tailoring the surface energy of carbon nanotubes mediates a macrophage immune response.

INTRODUCTION

Carbon nanotubes (CNTs) have attracted considerable attention due to their remarkable structural, electrical, and mechanical properties [1]. CNTs have been used to fabricate nanoscale drug delivery materials and sensors for the detection of proteins [2]. Individual CNTs as well as groups of CNTs have been used to change the surface energy of materials such as polymers [3] since CNTs increase surface roughness which is an important property for controlling surface energy. Such CNTs offer great opportunities for studying the interaction between the surfaces of cells and materials at the nanoscale level due to their intrinsic surface energy that has been related with cell functions such as adhesion, migration and proliferation [4]. According to a recent report [3], the change of surface energy in CNTs mixed with polymers promoted fibronectin absorption which is a key protein related to the function of numerous cells. Based on this fact, the surface of CNTs may also reduce immune responses by macrophages to promote implant function.

In this study, different macrophage responses on different surface energy CNTs were investigated. To determine the initial functions of macrophages on CNTs, the adhesion of macrophages was observed after a 4 hour seeding on both hydrophobic and hydrophilic carbon nanotubes.

EXPERIMENT

Hydrophobic and hydrophilic carbon nanotubes (each 5mg) were dissolved in chloroform (20ml) and ethanol (15ml), respectively. They were sonicated five times for 20 min with 6kW and hydrophobic and hydrophilic CNTs were placed on a disc (radius 18 cm), then dried in a vacuum oven to remove the toxicity of the solvent for 2 days. CNTs were exposed to UV light (264nm) for over 4 hours. Macrophages (TIB186 that are usually present after biomaterial implantation, obtained from ATCC) were cultured in RPMI 1640 medium (ATCC) with 10% (v/v) FBS and grown under 95% air and 5% CO_2. 5,000cells/cm^2 were seeded on CNT discs and the MTS assay was conducted to determine cell-viability after 4 hours. Experiments were repeated 6 times (N=6).

RESULTS AND DISCUSSION

CNTs discs were characterized by SEM, TEM and contact angles. SEM images (Fig. 1) showed the identical roughness of hydrophobic and hydrophilic CNTs. All hydrophobic and hydrophilic CNTs strongly adhered to the surface of a glass disc when flowing deionized water on their surfaces.

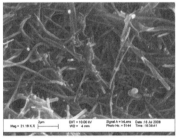   

Hydrophobic CNTs   Hydrophilic CNTs

Figure. 1 SEM images of hydrophobic and hydrophilic CNT surfaces on glass discs.

TEM characterization showed a difference between hydrophobic and hydrophilic CNT sizes (Fig. 2). Hydrophilic CNTs had a pyrolytic outer layer that was composed of oxides formed during the synthesis process.

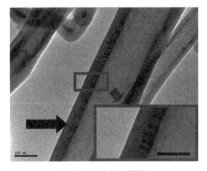   

(a) Hydrophilic CNTs   (b) Hydrophobic CNTs

Figure 2. TEM images of hydrophilic CNTs (a) and hydrophobic CNTs (b). The arrow in (a) indicates the pyrolytic outer layer. Each large box shows the magnification of the small rectangular area (scale bar in each box is 50 nm)

Such characteristics of the hydrophobic CNTs due to the pyrolytic layer decreased deionized water contact angles. Specifically, hydrophobic CNTs had a higher contact angle with deionized water (see Table 1) compared to hydrophilic CNTs (Fig. 3).

Table 1 Contact angles of hydrophobic CNTs and hydrophilic CNTs

Materials	Hydrophobic CNTs	Hydrophilic CNTs
Contact Angles (Tan value)	152°±1.4°	70°±3.5°

Hydrophobic CNTs   Hydrophilic CNTs

Figure 3. Contact angles on hydrophobic and hydrophilic CNTs.

Based on the above properties of hydrophobic and hydrophilic CNTs, results of this study showed that macrophages reacted differently to the four different surfaces (glass, titanium, hydrophobic and hydrophilic CNTs) after 4 hours (See Table 2).

Table 2 Macrophage cell adhesion after 4 hours

Materials	Glass	Titanium	Hydrophobic CNTs	Hydrophilic CNTs
Cells/cm^2	4,540±175	4,700±70	1,900±35	2,095±40

Interestingly, hydrophobic and hydrophilic CNTs which possessed the most rough surfaces, showed the least macrophage cell adhesion compared to the other substrates (glass and titanium). The increase of CNT roughness resulted in over two times less macrophage viability compared to titanium. This suggested that roughness in CNTs played a key role in reducing macrophage adhesion. In contrast, wettability of CNTs did not show an outstanding difference in the adhesion of macrophages as hydrophobic CNTs only slightly reduced the adhesion of macrophages. Therefore, in terms of macrophage adhesion, the results of this study showed that CNTs may be better candidates for bone implantation as they decreased the adhesion of a cell known to promote inflammation. For future studies, it is clear that not only number, but activation of macrophages, needs to be tested on these substrates.

CONCLUSIONS

For improving bone implants, inflammatory responses should be minimized. Based on the present study, CNTs may be a good candidate as a bone implant material due to not only their intrinsic mechanical and physical properties but also the presently observed relatively small adhesion of macrophages compared to glass or titanium.

ACKNOWLEDGMENTS

The authors would like to thank the Coulter and Hermann Foundation for funding.

REFERENCES

1. Lin Y, Taylor S, Li HP, Fernando KAS, Qu LW, Wang W, Gu LR, Zhou B, Sun YP. Advances toward bioapplications of carbon nanotubes. *J. Mater Chem*(2004)14;527-541
2. Chen RJ, Bangsaruntip S, Drouvalakis KA, Kam NWS, Shim M, Li YM, Kim W, Utz PJ, Dai HJ. Noncovalent functionalization of carbon nanotubes for highly specific electronic biosensors *PNAS*(2003)100(9);2984-4989
3. Khang D, Lu J, Yao C, Haberstroh KM, Webster TJ. The role of nanometer and sub-micron surface features on vascular and bone cell adhesion on titanium. *Biomaterials* (2008)29:8;970-983
4. Lu J, Rao MP, MacDonald NC, Khang D, Webster TJ. Improved endothelial cell adhesion and proliferation on patterned titanium surfaces with rationally designed micrometer to nanometer features. *Acta Biomater* (2008)4:1;192-201

Macromolecular Drug/Lipid Interaction

Mater. Res. Soc. Symp. Proc. Vol. 1140 © 2009 Materials Research Society　　　1140-HH02-02

Triterpene Saponin Glycosides: A New Class of Chemical Penetration Enhancers
Christopher Pino[1], V Prasad Shastri[1*]

1. Department of Biomedical Engineering, Vanderbilt University, Nashville, TN, USA 37235

* Corresponding Author:
Email: Prasad.Shastri@Vanderbilt.edu or Prasad.Shastri@gmail.com
Tel: 615-322-8005
Fax: 615-343-7919
Address: 5824 Stevenson Center, Station B #351631
Vanderbilt University, Nashville, TN 37235

ABSTRACT

The uppermost layer of the epidermis, the stratum corneum, provides the barrier properties to skin. Chemical penetration enhancers (CPEs) are chemicals capable of enhancing skin permeability to drugs. CPEs typically possess a lipid-like structure. Triterpene Saponin Glycosides (TSGs) are a class of planar macromolecules with ampiphilic characteristics, which have been shown to have therapeutic properties. The ampiphilic nature of TSGs makes them well-suited to mimic skin lipid components and hence good candidates for chemical penetration enhancement. Here we report that TSGs, whose molecular weights typically exceed the theoretical cut off for transdermal transport, are capable of providing enhancement of water soluble drugs through full thickness pig skin. Additionally, TSGs at higher concentrations are also capable of enhancing the transport of a hydrophobic model drug (estradiol). The mechanism of action is still under investigation, but preliminary data suggests that TSGs enhance transdermal transport through a combination of traditional pathways and secondary pathways that have not yet been fully explored.

INTRODUCTION

Transdermal delivery offers many advantages over other conventional forms of drug delivery such as oral administration, including an extended release profile, lower dosing requirements, enhanced patient compliance (Kligman 1984; Berti and Lipsky 1995). Transdermal delivery also bypasses first pass metabolism by the liver (Kogi, Tanaka et al. 1982), and allows for the direct application of drugs for localized delivery (Touitou, Meidan et al. 1998). However, transdermal delivery faces many hurdles associated with the barrier properties of the stratum corneum, the outermost layer of the skin (Guy, Hadgraft et al. 1987). The stratum corneum is made up of highly keratinized dead cell layers (Polakowska, Piacentini et al. 1994). These densely packed lipids give this barrier a very hydrophobic nature, with very limited permeability to water soluble molecules. This permeability can be modeled as the inverse square of molecular weight (Lian, Chen et al. 2008), with a cutoff of approximately 350 Daltons.

Hydrophilic, high mw drugs do not exhibit appreciable permeability across skin, and therefore many different methods of delivery have been studied to enhance transport. Mechanical disruption methods such as sonophoresis (Mitragotri, Edwards et al. 1995), electrical methods such as iontophoresis and electroporation (Prausnitz, Bose et al. 1993), and chemical penetration enhancers (Ranade 1991) have been used to deliver hydrophilic drugs. Because of their ease of formulation and application, CPEs are preferred over these other methods.

Chemical Penetration Enhancers, or CPEs can be grouped into chemical classes (Table 1): Surfactants, which can be anionic, cationic, Zwitterionic, or non-ionic surfactants, fatty acids and their derivatives, and azones and azone like molecules (Karande and Mitragotri 2003). Interestingly, all these CPEs have lipid like characteristics in that they are linear and possess a distinct hydrophobic region and a polar region. In addition these CPEs are only effective at high concentrations, and as a result elicit undesirable side effects at the site of application.

Triterpene Saponin Glycosides (TSGs) are naturally occurring glycosylated polycyclic hydrocarbons that have been shown to have many therapeutic uses. TSGs have been shown to possesses anti-oxidative, anti-inflammatory, anti-tumorigenic, anti-bacterial, and anti-fungal properties (Hanausek, Ganesh et al. 2001; Haridas, Arntzen et al. 2001; Haridas, Higuchi et al. 2001; Haridas, Hanausek et al. 2004; Haridas, Kim et al. 2005; Haridas, Li et al. 2007). TSGs are characterized by a planar triterpene core, with sugar side chains that can additionally contain monoterpene moieties (Jayatilake, Freeberg et al. 2003), and hence structurally differ vastly in comparison to conventional CPEs (Figure 1). It is important to point out that cyclic terpenes, that have previously been identified as CPEs are much lower in molecular weight in comparison to TSGs which range between 1000 - 2000 Daltons and have no structurally similarity to TSGs. Here we report the preliminary findings that suggest TSGs have potential as CPEs.

CPE class	Subclass	Example	Structure
Ionic Surfactant	Anionic	Sodium dodecyl sulfate (SDS)	
Ionic Surfactant	Cationic	Centromonium bromide	
Ionic Surfactant	Zwitterionic	Cocamidopropyl betaine	
Non-ionic Surfactant	Alkyl phenol poly(ethylene oxide)	Nonylphenol ethoxylates	
Non-ionic Surfactant	Fatty acid	Oleic acid	
Non-ionic Surfactant	Fatty alcohol	Oleyl alcohol	
Azones		Hexyl azone	
Planar, Amphipathic Macromolecules		Triterpene Saponin Glycosides	

Table 1: Classes of Chemical Penetration Enhancers (CPEs) with example chemical structures.

METHODS

Full Thickness Pigskin Transport Experiments

Pigskin from the back region of adult female pigs was obtained from Lampire Biological Laboratories (Pipersville, PA). Skin was thawed overnight at 5°C, cut into 1 inch x 1 inch pieces, and mounted in side-by-side glass diffusion cells with an inner diameter of 5 mm (Permegear, Hellertown, PA) with the stratum corneum facing the donor compartment. Both donor and receiver compartments were filled with 2 mL PBS and hydrated for 2 hours prior to any experiment. Barrier integrity was verified by measuring AC electrical conductance across the skin at 1 kHz, 143.0 mV signal amplitude (Agilent 33220A Function Generator). Skin pieces with conductance values ranging between 8-20 µA were used in the study (Lee, Langer et al. 2003). To start transport experiments, PBS was removed from the donor compartment and was replaced with 2 mL of the transdermal formulation with the drug of interest and the receiver compartment was filled with fresh PBS. All studies were carried out at in triplicate, at an ambient temperature of 25° C, as regulated and measured by the thermostat in the laboratory. Both the donor and receiver compartments were stirred continuously using microflea stir bars, throughout the duration of the study. At pre-determined time points of 3, 6, 12, 18, and 24 hours, the solution from receiver compartment was collected for analysis and replaced with 2 mL of fresh PBS. After the final time point, integrity of the skin sample was once again verified using electrical measurements.

HPLC Analysis of bupivacaine transport

Bupivacaine HCl was quantified using by HPLC (Shimadzu HPLC System, SCL-10A System Controller, dual LC10AD-VP pumps, DGU-14A degasser, FCV-10AL mixer, SPD-10MA Diode Array Detector, SIL-10AD Auto Injector) through a reverse-phase column (Waters C_{18} uBondaPak^{TM} 3.9x300 mm). Each sample was filtered using a 0.45 µm PTFE filter before analysis. The mobile phase used was DI H_2O (5% acetic acid, 2% 5M sodium hydroxide; pH = 4.2)/acetonitrile (35:65 v/v). Bupivacaine was detected at 237 nm under isocratic flow conditions (flow rate = 1.6mL/min). Calibration curves were generated using bupivacaine dilutions. The retention time for bupivacaine was 4.9 – 5.6 minutes.

Scintillation analysis of 3-H Estradiol

Each 2ml receiver compartment sample at each time point was transferred into a scintillation vial and 2 ml of scintillation fluid was added. Each sample was run through a Beckman LS6500 beta counter, and raw scintillation counts of radioactivity were printed out. Known concentrations of ^{3}H-estradiol was used to create standard concentration curves to correlate radioactive measurements to mass. Data analysis was done in Excel, and data was reported as averages +/- standard deviation to represent all replicates. Flux and permeability values were calculated based on values within steady state transport regions.

RESULTS AND DISCUSSION

TSGs enhance the transdermal transport of water soluble anesthetics

As an example we studied bupivicane HCl, a water soluble anesthetic, with a molecular weight approaching the theoretical mw cutoff for skin, and having a moderate octanol/water partition coefficient of 1.7. We found that the addition of TSGs increased the cumulative amount of bupivacaine delivered across full thickness pig skin after 24 hours (Figure 1, Table 2). Additionally, almost an order of magnitude increase in permeability at steady-state was observed (Table 2). This demonstrates that TSGs are capable of enhancing water

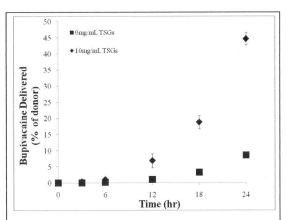

Figure 1. Bupivacaine HCl transport in presence and absence of TSGs.

soluble drug delivery even at low concentrations of 1% w/v with an observed enhancement of 4.5 fold at 24 hours. We have observed similar trends for all other water soluble anesthetics (Pino 2008) and water soluble cardiovascular drugs (ibid).

TSGs impact estradiol transport through full thickness pig skin

To elucidate the mechanism of action of TSGs, we assessed the transport of estradiol, whose transport is believed to occur primarily through lipid pathways. An enhancement of estradiol delivery when administered with TSGs would likely indicate that TSGs are capable of affecting lipid packing. Estradiol transport across full thickness pigskin is shown in Figure 2. At low concentrations of TSGs, there doesn't appear to be a significant difference between permeability of skin to estradiol with or without TSGs (data not shown), however, at 20 mg/ml of TSGs, an appreciable increase in

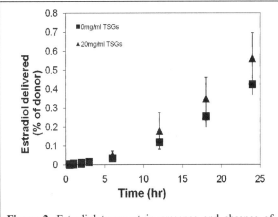

Figure 2. Estradiol transport in presence and absence of TSGs.

estradiol transport at 24 hours was observed (Figure 2, Table 2). The modest increase in permeability of estradiol (Table 2) suggests that although TSGs appear to affect lipid pathways, this effect at best is nominal.

TSG concentrations	**0 mg/ml TSG**	**10 mg/ml TSG**	**Enhancement**
Bupivacaine Permeability (cm/hr)	7.08×10^{-4}	3.17×10^{-3}	347%

TSG concentrations	**0 mg/ml TSG**	**20 mg/ml TSG**	**Enhancement**
Estradiol Permeability (cm/hr)	2.76×10^{-4}	3.62×10^{-4}	31%

Table 2. Transport enhancement of bupivacaine HCl and estradiol across full thickness pigskin, in the presence of TSGs.

Possible mechanisms of TSG action

Conventional CPEs typically act through the disruption of lipids, as their linear structure allows them to intercalate within lipid bilayers. In addition, conventional CPEs also enhance transport of lipophilic drugs by increasing the solubility of the drug in lipids (Thong, Zhai et al. 2007). The transport behavior of estradiol and bupivacaine HCl taken in sum provide some insight into the possible pathways TSGs affect in skin. The increase in estradiol transport suggests that TSGs may associate with lipids to some extent, especially at high concentrations. The finding that skin resistivity measurements are not statistically different, before and after TSG application, suggests that skin barrier properties are maintained. The trends observed in the bupivacaine system and other water soluble drug studies imply that secondary mechanisms of transport involving aqueous pathways may be at play as well (Pino 2008)

CONCLUSIONS

TSGs, which are planar macromolecules, represent a new class of amphipathic CPEs. TSGs act as enhancers for both hydrophilic and hydrophobic model drugs at different concentration regimens. TSGs exhibit appreciable CPE action even at concentrations as low as 1% w/v. The mechanism of action is still under investigation, but preliminary data suggests that TSGs enhance transdermal transport through a combination of traditional pathways and secondary pathways that have not been fully explored.

ACKNOWLEDGMENTS

We'd like to thank Michael Scherer for his help in developing HPLC protocols and for his aid in completing some of the bupivacaine transport experiments. We'd also like to acknowledge the Clayton Foundation for Research for their generous support of this work.

REFERENCES

Berti, J. J. and J. J. Lipsky (1995). "Transcutaneous drug delivery: a practical review." Mayo Clin Proc **70**(6): 581-6.

Guy, R. H., J. Hadgraft, et al. (1987). "Transdermal drug delivery and cutaneous metabolism." Xenobiotica **17**(3): 325-43.

Hanausek, M., P. Ganesh, et al. (2001). "Avicins, a family of triterpenoid saponins from Acacia victoriae (Bentham), suppress H-ras mutations and aneuploidy in a murine skin carcinogenesis model." Proc Natl Acad Sci U S A **98**(20): 11551-6.

Haridas, V., C. J. Arntzen, et al. (2001). "Avicins, a family of triterpenoid saponins from Acacia victoriae (Bentham), inhibit activation of nuclear factor-kappaB by inhibiting both its nuclear localization and ability to bind DNA." Proc Natl Acad Sci U S A **98**(20): 11557-62.

Haridas, V., M. Hanausek, et al. (2004). "Triterpenoid electrophiles (avicins) activate the innate stress response by redox regulation of a gene battery." J Clin Invest **113**(1): 65-73.

Haridas, V., M. Higuchi, et al. (2001). "Avicins: triterpenoid saponins from Acacia victoriae (Bentham) induce apoptosis by mitochondrial perturbation." Proc Natl Acad Sci U S A **98**(10): 5821-6.

Haridas, V., S. O. Kim, et al. (2005). "Avicinylation (thioesterification): a protein modification that can regulate the response to oxidative and nitrosative stress." Proc Natl Acad Sci U S A **102**(29): 10088-93.

Haridas, V., X. Li, et al. (2007). "Avicins, a novel plant-derived metabolite lowers energy metabolism in tumor cells by targeting the outer mitochondrial membrane." Mitochondrion **7**(3): 234-40.

Jayatilake, G. S., D. R. Freeberg, et al. (2003). "Isolation and structures of avicins D and G: in vitro tumor-inhibitory saponins derived from Acacia victoriae." J Nat Prod **66**(6): 779-83.

Karande, P. and S. Mitragotri (2003). "Dependence of skin permeability on contact area." Pharm Res **20**(2): 257-63.

Kligman, A. M. (1984). "Skin permeability: dermatologic aspects of transdermal drug delivery." Am Heart J **108**(1): 200-6.

Kogi, K., O. Tanaka, et al. (1982). "[Hemodynamic effects of a transdermal formulation of isosorbide dinitrate and its pharmacokinetics in conscious dogs]." Nippon Yakurigaku Zasshi **80**(4): 279-88.

Lee, P. J., R. Langer, et al. (2003). "Novel microemulsion enhancer formulation for simultaneous transdermal delivery of hydrophilic and hydrophobic drugs." Pharm Res **20**(2): 264-9.

Lian, G., L. Chen, et al. (2008). "An evaluation of mathematical models for predicting skin permeability." J Pharm Sci **97**(1): 584-98.

Mitragotri, S., D. A. Edwards, et al. (1995). "A mechanistic study of ultrasonically-enhanced transdermal drug delivery." J Pharm Sci **84**(6): 697-706.

Pino, C., Scherer, MA, Shastri VP (2008). "Percutaneous Delivery of Water Soluble Anesthetics From Triterpene Saponin Glycoside Formulations." J Pharm Sci.

Polakowska, R. R., M. Piacentini, et al. (1994). "Apoptosis in human skin development: morphogenesis, periderm, and stem cells." Dev Dyn **199**(3): 176-88.

Prausnitz, M. R., V. G. Bose, et al. (1993). "Electroporation of mammalian skin: a mechanism to enhance transdermal drug delivery." Proc Natl Acad Sci U S A **90**(22): 10504-8.

Ranade, V. V. (1991). "Drug delivery systems. 6. Transdermal drug delivery." J Clin Pharmacol **31**(5): 401-18.

Thong, H. Y., H. Zhai, et al. (2007). "Percutaneous penetration enhancers: an overview." Skin Pharmacol Physiol **20**(6): 272-82.

Touitou, E., V. M. Meidan, et al. (1998). "Methods for quantitative determination of drug localized in the skin." J Control Release **56**(1-3): 7-21.

Mater. Res. Soc. Symp. Proc. Vol. 1140 © 2009 Materials Research Society 1140-HH09-07

Computational modeling of anionic and zwitterionic lipid bilayers for investigating surface activities of bioactive molecules

Jutarat Pimthon[1,2], Regine Willumeit[3], Andreas Lendlein[1], and Dieter Hofmann[1]*

[1]Institute of Polymer Research, GKSS Research Center, Kantstr. 55, D-14513 Teltow, Germany
[2]Department of Pharmaceutical Chemistry, Faculty of Pharmacy, Mahidol University, 447 Sri-Ayudhya Road, Bangkok 10400, Thailand.
[3]Institute of Materials Research, GKSS Research Center, Max-Planck-Str., D-21502 Geesthacht, Germany

ABSTRACT

Phosphatidylgylcerols (PGs) and Phosphatidylethanolamines (PEs) are of considerable interest because they are major lipid components of bacterial membranes and the non-charged PE can also serve as a model for cell membranes of multi-cellular organisms. Here, we report molecular dynamics (MD) simulations studies of the structural and dynamics properties of negatively charged POPG and zwitterionic POPE bilayers. The hydrocarbon chain fluidity and the electron density distribution of various groups along the bilayer were extensively analyzed and compared with the available experimental data. A specific focus was given on hydrogen bond formations and position of sodium ions in the lipid bilayers. These validated and equilibrated models were subsequently employed to investigate selectivity and mechanism of action of the antimicrobial peptide of interest. We found that hydrogen bonding and electrostatic interactions potentially play a role in the adsorption of a peptide to the membrane interface. We observed the peptide's insertion into the membrane can decrease the order parameter and induce local membrane deformation.

INTRODUCTION

Biological cell membranes are complex multicomponent assemblies. Phospholipids, however, function as major structural elements of biological membranes, in which their physical characteristics are key determinants of membrane structure and function (for recent review see [1]). Due to fluctuations in the biologically relevant lipid fluid phase (L_c) however, experimental, difficulties are encountered in obtaining models of membranes at atomistic detail [2]. It has been shown that atomic-scale computer simulation can provide powerful assistance in understanding structural and dynamical properties of lipid membrane that are inaccessible experimentally (for review see [3-5]).

The main goal of the present study is to characterize by atomistic MD simulations the structural and dynamical properties of fluid phase lipid bilayers of 1-palmitoyl-2-oleoyl-phosphatidyl-glycerol (POPG) and 1-palmitoyl-2-oleoyl-phosphatidylethanolamine (POPE), which are the most widely utilized as model systems for mimicking cytoplasmic bacterial membranes. Structures of the phospholipids used in this study are shown in Figure 1. One of the striking differences between PG and PE headgroups is that PG has a net negative charge at physiological pH, whereas the zwitterionic PE carries no net charge. A more fundamental understanding of those phospholipid model membranes will not only be suited for studying the organization and dynamics of the differently charged phospholipids in model membranes but these models can also help to investigate the mechanism of interactions with antibacterial compounds.

As part of our ongoing research on peptide–membrane interactions, the initial step involved in the peptide binding and selectivity at the biomembrane interfaces is a possible key process. In this study we focused on the mode of interaction between the synthetic antimicrobial peptide NK-2 ([6] and reference therein) and a model membrane of pure POPG bilayer, and a POPE bilayer. The experience gained from these studies will shape the direction of tracing surface immobilization and activity of bioactive peptides on biomembranes using computer simulation techniques.

COMPUTATIONAL DETAILS

All simulations and analyses were performed using the CHARMM package [7] with the all atom PARAM27 force field [8]. Simulations of bilayers carried out in the study were: (i) a negatively charged POPG bilayer consisting of 224 POPG molecules, 224 sodium ions, and 5589 water molecules, (ii) a zwitterionic POPE bilayer consisting of 224 POPE molecules and 5229 water molecules. Details of the model will be presented elsewhere. In brief, the starting POPG (16:0-18:1) configuration was built based on the structure of diplamitoylphosphatidylglycerol (DPPG, 16:0/16:0-PG) membrane [9]. For POPE bilayer, the starting structure was built by modification of a diplamitoylphosphatidylethanolamine (DPPE, 16:0/16:0-PE) [9]. The preceding bilayer simulations were carried out at NPAT ensemble, where the area per the lipid was about 62 Å^2 [10] and 59 Å^2 [11] for POPG and POPE bilayers, respectively. For peptide-membrane association models, the systems simulated contained (i) 6 NK-2 peptide, 224 POPG lipids, 16649 water molecules, 216 sodium ions, and 52 chloride ions (ii) 6 NK-2 peptide, 224 POPE lipids, 14587 water molecules, 48 sodium ions, and 108 chloride ions. More details of the model set-up can be found else where. The pure lipid simulations were carried out for 40 ns and the peptide-lipid model system were carried for 6 ns in periodic boundary conditions.

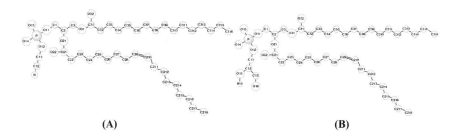

(A)   (B)

Figure 1. Chemical structures and the numbering of atoms of POPG (A) and POPE (B) molecules used in this study

RESULTS AND DISCUSSION

General structure of the lipid bilayer

To characterize the global structure of the liquid crystalline phase of POPG and POPE, the electron density profiles of selected groups [12] and acyl chain order parameters [13] were calculated at the fixed surface density based on the experimental area per lipid at 30 °C. Figure 2A and 2B show the average electron density profiles of the POPG and POPG bilayer, respectively. For the charged POPG bilayer, we are mainly interested in the distribution of the sodium ions. Based on the profile (Figure 2A), we observed the tendency of sodium ions to reside close to head groups as its distribution at the interface region overlaps with the peaks of the glycerol headgroup and phosphodiester groups. This is consistent with the simulated fluid phase DPPG bilayer [14]. In the zwitterionic POPE membrane (Figure 2B), the profile indicates that the amine group prefers to interact (hydrogen bond formation) with the phosphate lipid head group as seen by about the same peak positions of those groups in the profile. This indicated the characteristic biophysical property of preferred hydrogen-bonding capability of PE polar headgroups. The simulated phosphate-phosphate distance of POPG is around 40.7 Å. In this case no experimental data is available. However, this feature is in good agreement with the previous simulation by Zhao *et. al.* [15]. For POPE on the other hand, we obtained the P-P distance at about 40 Å which is consistent with available experiment data [11].

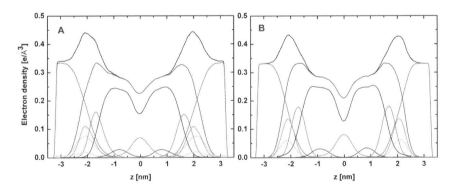

Figure 2. The average electron density profiles for various lipid functional groups along the normal to the bilayer surface (z axis) of POPG system (A), and POPE system (B). Center of membrane is located at $z=0$. Shown are the profiles for total (——), lipid (——), water (——), ethanolamine (——), glycerol headgroup (——), phosphate (——), glycerol backbone (——), methylene (——), double bond (——), methyl (——), sodium ions (——).

The deuterium order parameters of lipid tails, $|S_{CD}|$, give a measure of the orientation and ordering of the phospholipid tails in the bilayer with respect to the bilayer normal. Note that a

$|S_{CD}|$ value of 0.5 corresponds to perfect alignment of the lipid tail to the normal of the bilayer surface. Figure 3 shows the order parameters of both systems. In addition to there is available experimental data for orientational order of acyl chain of POPC [16], it has the same hydrocarbon chain core as POPG and POPE. Such that. the selectively deuterated hydrocarbon chains of POPC bilayer (black triangle) from literature data are also shown in the Figure 3 for comparison and validate the simulation models. We observed that the overall $|S_{CD}|$ profile of POPG (black line) indicates more disorder than in the case of POPE (dotted grey line), due to the fact that the effective size of the PE polar headgroup is smaller. For the oleoyl profile (Figure 3A), the simulation data of both systems are consistent with experimental POPC data, in which a disordered trend around the double bond is indicated. In addition, the region near the double bond is in relatively good agreement with experiment. The other characteristic feature is the splitting at carbon 2 on oleoyl chain. POPE shows excellent agreement with experiments, while such a non-equivalent methylene proton was not observed in POPG model. For the palmitoyl chain, shown in Figure 3B, there is a significant departure from experiment of $|S_{CD}|$ at position C2 in both systems, in which it is found to be more disordered than the selectively deuterated C2 position of POPC. Unexpectedly, the simulated $|S_{CD}|$ value of palmitoyl position C2-4 of POPE is also smaller than that observed experimentally for POPG and POPC. This may be explained by the fact that the mobility of the headgroups is rather slow in this case which might require rather long simulation times to get quantitative agreement [17, 18]. However, the qualitative agreement of the shape of the entire profile among all investigated models is acceptable. A more detailed account of the results presented here will be given elsewhere.

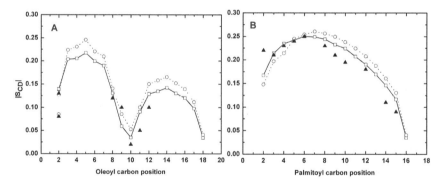

Figure 3. Comparison of oleoyl (A) and palmitoyl chain (B) simulated order parameter profile of POPG (—□—) and POPE (⋯○⋯). For comparison, experimental deuterated POPC (▲) order parameters are shown [16].

Peptide-membrane association

In our previous studies, it was demonstrated how the NK-2 induced a local deformation of the gel phase bilayer model membrane (data not shown). Here we continue our work to investigate its interaction with liquid crystalline phase model membranes and at higher concentrations of the peptides. As indicated by energy calculation (data not shown), the electrostatic interactions including hydrogen bond formation play a critical role in membrane association of NK-2. Moreover, due to the less energetically favorable situation, such interaction is less pronounced in the POPE membrane (data not shown). This effect is clearly seen in Figure 4A and 4B that the peptide shows a stronger tendency to bind to the PG than to the PE during the same period of simulation time. After adsorption of the peptide to the membrane surface, in particular the change in lipid membrane morphology was monitored by calculating chain orientational order parameters. The differences of peptide binding on the two different types of bilayers (only monolayer in contact with peptides considered) is shown in Figure 4C-4F. In the peptide-POPG model, (Figure 4C, 4D), we found a substantial decrease in order parameters of the lipid chain for the lipids that are in the peptide-bound monolayer. While, there are no pronounced effects on the peptide-free monolayer and peptide-bound monolayer in the peptide-POPE system (Figure EC, 4F). This confirms the selectivity and strong binding of the peptide to the negatively charged bacterial membrane over the mammalian cell membrane. A more detailed analysis of the mechanism of membrane perturbation by NK-2 will be reported elsewhere.

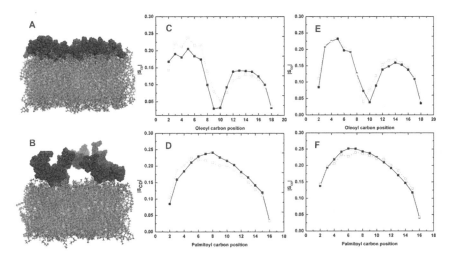

Figure 4: Snapshoot of the 6-ns of; (A) NK-2/POPG system, (B) NK-2/POPE. Lipid order parameters of; (C) the oleoyl chain of POPG, (D) the palmitoyl of POPG, (E) the oleoyl chain of POPE, (D) the palmitoyl of POPE. The order parameters are plotted for lipids in the peptide-free monolayer ($\cdots\square\cdots$) and in the monolayer, in which the peptide bind ($-\square-$).

CONCLUSIONS

Computational models for membranes were set up and utilized to represent the phospholipid bilayer domain of bacterial cell membranes and mammalian cell membranes. The orientational order profiles of the lipid chains were calculated to monitor the fluidity of the model membrane. We found a good agreement with experimental data. The electron density profiles of various functional groups along the bilayer normal indicate that the sodium ions reside close to the POPG membrane surface and bind strongly with the phosphate oxygen. This is consistent with other simulations [14]. The fact that the ethanolamine and phosphate of POPE headgroup occupy about the same area indicates the intra hydrogen bond formation, a characteristic of phosphatidylethanolamine. As the main observations were in line with the experiments, it can be concluded that our computational model membrane can serve as a model system for many biophysical applications. Here one specific application was the use of such models to understand the selectivity of peptide binding on the membrane surface. We observed that the charged residues of peptides form a hydrogen bond with the lipid head group thus stabilizing the peptide-membrane interaction. Consequently, respective effects on the fluidity of membrane model are observed. Further deep analysis will be performed to understand the mode of interaction of the peptide on the surface as such can help to improve the molecular design e.g., of new potent antimicrobial compounds.

ACKNOWLEDGMENTS

J.P. acknowledges support from The Royal Thai Government.

REFERENCES

1. G. van Meer, D. R. Voelker and G. W. Feigenson, *Nat. Rev. Mol. Cell. Biol.* **9.** 112 (2008).
2. J. F. Nagle and S. Tristram-Nagle, *Biochim. Biophys. Acta.* **1469.** 159 (2000).
3. S. E. Feller, in *Methods in Molecular Biology (Methods in Membrane Lipids)*, vol. **400**, edited by A. M. Dopico (Humana Press, 2007), p. 89.
4. R. W. Pastor, R. M. Venable and S. E. Feller, *Acc. Chem. Res.* **35.** 438 (2002).
5. M. Karttunen, J. Rottler, I Vattulainen, and C. Sagui in *Current topics in membranes; Computational Modeling of Membrane Bilayers*, vol. **60**, edited by S. Feller (Elsevier, 2008), p. 49.
6. R. Willumeit, M. Kumpugdee, S. S. Funari, K. Lohner, B. P. Navas, K. Brandenburg, S. Linser and J. Andra, *Biochim. Biophys. Acta.* **1669**, 125 (2005).
7. B. R. Brooks, R. E. Bruccoleri, B. D. Olafson, D. J. States, S. Swaminathan and M. Karplus, *J. Comput. Chem.* **4,** 187 (1983).
8. S. E. Feller, and A. D. MacKerell, *J. Phys. Chem. B.* **104,** 7510 (2000).
9. J. Pimthon, R. Willumeit, A. Lendlein, D. Hofmann, *J. Mol. Struct*, in press.
10 G. Pabst. (personal communication).
11. M. Rappolt, A. Hickel, F. Bringezu and K. Lohner, *Biophys. J.* **84,** 3111(2003).
12. S. E. Feller, R. M. Venable and R. W. Pastor, *Langmuir* **13,** 6555 (1997).
13. L. Vermeer, B. de Groot, V. Réat, A. Milon and J. Czaplicki, *Eur. Biophys. J.* **36**, 919 (2007).
14 T. Zaraiskaya and K. R. Jeffrey, *Biophys. J.* **88,** 4017 (2005).
15. W. Zhao, T. Róg, A. A. Gurtovenko, I. Vattulainen, and M. Karttunen, *Biophys. J.* **92,** 1114 (2007).
16. A. Seelig and J. Seelig, *Biochemistry* **16**, 45 (1977).
17. F. Suits, M. C. Pitman and S. E. Feller, *J. Chem. Phys.* **122**, 244714 (2005).
18. H. Martinez-Seara, T. Róg, M. Pasenkiewicz-Gierula, I. Vattulainen, M. Karttunen, and R. Reigada, *J. Phys. Chem. B.* **111**, 11162 (2007)

Mater. Res. Soc. Symp. Proc. Vol. 1140 © 2009 Materials Research Society   1140-HH05-32

Triterpene Saponin Glycosides impact the percutaneous delivery of water soluble drugs

Christopher J. Pino[1], Michael A. Scherer[1], C. Monica Guan[1] and V. Prasad Shastri[1*]

1. Department of Biomedical Engineering, Vanderbilt University, Nashville, TN, USA 37235

* Corresponding Author:
Email: Prasad.Shastri@gmail.com
Tel: 615-322-8005
Fax: 615-343-7919
Address: 5824 Stevenson Center, Station B #351631
Vanderbilt University, Nashville, TN 37235

ABSTRACT

Percutaneous absorption and transdermal delivery of water soluble drugs have proven to be challenging due to (a) their poor partitioning into lipids and (b) the limited potential of traditional lipophilic chemical penetrations enhancers (CPEs) in providing appreciable transport. To date enhancement from water-based systems has been limited to ethanol/water, water/N-methyl pyrrolidone, binary systems with or without augmentation with surfactants. In an ongoing study, we have shown that triterpene saponin glycosides (TSGs) can significantly impact barrier properties of skin. In this study we evaluated the effect of TSGs on the transport of water-soluble cardiovascular drugs currently in clinical use across full thickness pig skin. TSGs were found to enhance the permeability of these drugs, by a fold factor of 1.4 – 10.1. Surprisingly, the observed enhancement did not inversely correlate with molecular weight.

INTRODUCTION

A majority of medications contain water soluble actives. Among cardiovascular drugs, the most prescribed, antihypertensive drugs, are typically water soluble, and have appreciable molecular weight. These properties preclude these drugs from efficacious administration via transdermal delivery (Guy et al. 1987; Guy and Hadgraft 1988). Despite the advantages of transdermal delivery which include increased patient compliance, reduced first-pass effect and localization of therapy (Sclar et al. 1991; Prisant et al. 1992; Cramer and Saks 1994), most hypertensive drugs continue to be administered orally because of their limited permeation across skin.

Due to the limited percutaneous absorption of water soluble drugs, strategies to enhance transdermal delivery have relied on mechanical disruption of the skin barrier using sonophoresis (Mitragotri et al. 1995), iontophoresis (Singh and Roberts 1989), or electroporation (Prausnitz et al. 1993). More recently, it has been shown that with appropriate combination of chemical penetration enhancers (CPEs) some peptides may be delivered transdermally (Karande et al. 2004). CPEs offer several advantages over mechanical disruption of stratum corneum (SC) including reversible permeabilization of the SC barrier, relative ease of formulation, and application. To date, enhancement from water based systems has been limited to ethanol/water

(Kobayashi et al. 1993; Lee et al. 1994), water/N-methyl pyrrolidone (NMP) (Lee et al. 2005), and binary systems with or without surfactants. These systems have some inherent disadvantages in that they require a high concentration of an organic vehicle to carry the CPE, which can lead to lipid extraction, and skin irritation, thereby precluding repeated administration of the delivery system at the site of application. As such, due to their lipophilicity, CPEs are much better suited for delivery of hydrophobic molecules. A water soluble CPE on the other hand, will allow for an aqueous formulation, which in addition to having better dermatological profile, may also enable the transdermal delivery of water-soluble drugs.

Triterpene Saponin Glycosides (TSGs) are water-soluble natural products extracted from the desert plant *Acacia Victorae*, with amphiphilic characteristics, that behave as CPEs (Pino 2008). In this study we investigated the potential of TSGs to enhance the transdermal transport of various hypertensive drugs: α-Propranolol Hydrochloride (HCl), Losartan Potassium (K), and Diltiazem Hydrochloride (Table 1).

METHODS

Transdermal experiments

Transport of antihypertensive drugs was studied using full thickness pig skin from the dorsal region of adult female pigs which was obtained from Lampire Biological Laboratories (Pipersville, PA), and stored at -80 °C. Skin was thawed overnight at 5°C, cut into 1 inch x 1 inch pieces, and mounted in side-by-side glass diffusion cells with an inner diameter of 10 mm (Permegear, Hellertown, PA) with the stratum corneum facing the donor compartment. Both donor and receiver compartments were filled with 2 mL PBS and hydrated for 2 hours prior to any experiment. Barrier integrity was verified by measuring AC electrical conductance across the skin at 1 kHz, 143.0 mV signal amplitude (Agilent 33220A Function Generator). Skin pieces with conductance values ranging between 8-20 µA were used in the study. At the commencement of the transport experiments, PBS was removed from the donor compartment and replaced with 2 mL of the aqueous solution of drug with or without TSGs, and the receiver compartment was filled with fresh PBS. All studies were carried out at in triplicate, at an ambient temperature (25° C). Both the donor and receiver compartments were stirred continuously using microflea stir bars, throughout the duration of the study. At pre-determined time points of 3, 6, 12, 18, and 24 hours, the solution from receiver compartment was collected for analysis and replaced with 2 mL of fresh PBS. After the final time point, integrity of the skin sample was once again verified using electrical measurements.

Drug	α-Propranolol-HCl	Diltiazem-HCl	Losartan-K
Class	β-blocker, non specific β-adrenergic agonist	Calcium channel blocker	Angiotensin-II receptor agonist
Structure			
Molecular Weight (Da)	295.8	414.5	423
$logP_{ow}$ at pH 7	1.2	2.89	1.19
$t_{1/2}$ in Plasma (hrs)	4 - 5	3 - 4.5	1.5 - 2
Aqueous Solubility	Moderate-High	High	High

Table 1: Properties of water soluble antihypertensive drugs studied.

HPLC analysis of cardiovascular drugs

Losartan K and α-propranolol HCl were quantified by reverse phase HPLC (Shimadzu HPLC System, SCL-10A System Controller, dual LC10AD-VP pumps, DGU-14A degasser, FCV-10AL mixer, SPD-10MA Diode Array Detector, SIL-10AD Auto Injector). Each sample was filtered using a 0.45 μm PTFE filter before analysis. Propranolol was analyzed using a C8 column (Agilent Eclipse plus C8, 5μm, 4.6 x 150 mm), an aqueous mobile phase of [0.15M phosphoric acid, 0.25% sodium dodecyl sulfate (SDS)]/acetonitrile/methanol (10:45:45 v/v) under isocratic flow conditions (flow rate = 1.5 mL/min), and detected at 280 nm. For losartan potassium, a C18 column was used (Agilent Eclipse plus C18, 5μm, 4.6 x 150 mm), with a mobile phase of (60:40) 0.1% phosphoric acid in water: acetonitrile, at a flow rate of 1.5 mL/min, and detected at 254 nm. The retention time for propranolol HCl was 3.4 – 3.9 minutes, and losartan K was 2.3 - 2.8 min.

Scintillation analysis of ^{3}H-Diltiazem and ^{3}H-Propranolol

When radiolabeled drugs were used, at every sampling time point, the receiver compartment volume (2 ml) was removed for analysis, and replaced with fresh PBS. The sample was then mixed with 2ml of scintillation fluid (Atlantic Nuclear). Each sample was analyzed using a Beckman LS6500 beta counter. Standard concentration curves were generated using dilutions of known concentrations of ^{3}H-Diltiazem and ^{3}H-Propranolol, to correlate radioactive measurements to mass. Data analysis was done in Excel, and data was reported as averages +/- standard deviation to represent all replicates. Flux and permeability values were calculated based on values within steady state transport regions.

RESULTS AND DISCUSSION

TSGs enhance the transdermal transport of water soluble cardiovascular drugs

Recently we reported that TSGs are capable of providing significant transdermal enhancement of water soluble forms of local anesthetics (Pino 2008). These enhancing effects were observed at concentrations of TSGs as low as 0.1% w/v. In this study, we evaluated the use of TSGs to enhance the transport of antihypertensive drugs across full thickness pig skin. The salient findings of this study are presented in this report. TSGs, at a donor concentration of 1% w/v, afforded a significant enhancement in transdermal transport of all three drugs: diltiazem HC, losartan K, and α-propranolol HCl. In all cases, the cumulative mass transport beyond 6 hours was significantly greater from formulations with TSGs.

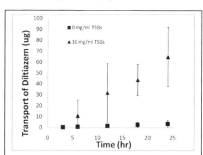

Figure 1: Enhancement of transdermal transport of Diltiazem HCl from formulations with and without TSGs across full thickness pig skin.

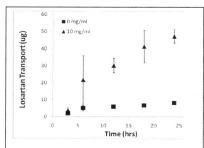

Figure 2: Transport of losartan K from aqueous formulations with and without TSGs across full thickness pig skin.

In Figure 1, the transport of diltiazem HCl across full thickness pig skin is shown. In presence of TSGs, nearly a 10 fold increase in cumulative transport at 24 hours was observed, which was accompanied by an order of magnitude increase in permeability (Table 2). These transport characteristics are unexpected considering that the molecular weight of diltiazem HCl is higher than the theoretical molecular weight cut off for skin.

The transport of losartan K across full thickness pig skin was also enhanced by the inclusion of TSGs in aqueous formulations as shown in Figure 2. Although, in comparison with diltiazem, the fold enhancement of cumulative losartan K transport at 24 hours in presence of

TSGs was lower, this increase in permeability was still significant and was nearly 8 fold greater than controls. Standard deviations of these transport studies are high, which is not uncommon in the transdermal field due the inherent variability of skin samples.

Propranolol-HCl transport on the other had is not impacted by the addition of TSGs (Figure 3, Table 2). The enhancement in permeability was modest around 40% in comparison controls. An important commonality among the drugs investigated in this study is their high water solubility. In view of this fact, it appears that when solubility differences are minimal, the enhancement of drugs in presence of TSGs appear to follow the trend of increasing lipophilicity. Diltiazem which has the highest octanol-water partition coefficient ($\log P_{ow}$) benefited most for the inclusion of TSGs in the formulation.

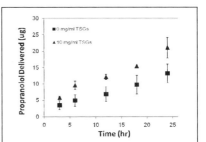

Figure 3: Transport of propranolol HCl from aqueous formulations with and without TSGs across full thickness pig skin

Drug	Permeability (cm/hr)	Permeability with 1% (w/v) TSGs	Enhancement factor
α-Propranolol HCl	3.35×10^{-4}	4.74×10^{-4}	1.4
Losartan K	1.15×10^{-4}	9.02×10^{-4}	7.8
Diltiazem HCl	9.32×10^{-5}	9.44×10^{-4}	10.1

Table 2: Permeability of cardiovascular drugs across skin in presence and absence of TSGs: Enhancement factor was calculated as the ratio of the permeability of the drug in the aqueous solutions containing TSGs divided by the permeability of the drug in formulations without TSGs.

Differences in cardiovascular drug transport enhancement

One can infer a bit about the mechanism by which TSGs impact drug transport. When propranolol HCl transport enhancement is compared with losartan K, which both have the similar $\log P_{ow}$, it is clear that the higher MW drug, losartan benefits much more from administration from a TSG formulation, as demonstrated by the enhancement factor. The findings of this work are consistent with our observations in an earlier study. We have previously shown that transport enhancement of water soluble anesthetics by TSGs did not necessarily correlate with the inverse of the MW of the anesthetic delivered (Pino et al 2008). When one compares transport characteristics of diltiazem HCl and losartan K, it is clear that when the MW of the drug is greater than the theoretical cutoff, lipophilicity dominates. Diltiazem, which is the more hydrophobic of the two by (higher $\log P_{ow}$) benefits more from the introduction of TSGs. This data taken in sum suggests that the mechanism of action by TSGs involve interactions with lipids.

CONCLUSIONS

We have shown that TSGs can enhance the transdermal delivery of three of the most commonly prescribed antihypertensive drugs. This is consistent with our earlier finding with water soluble local anesthetics. The identification of efficacious water soluble CPEs should prove to be helpful in the development of transdermal systems for the delivery of cardiovascular drugs.

ACKNOWLEDGMENTS

This work was supported by the Clayton Foundation for Research.

REFERENCES

Cramer, M. P. and S. R. Saks (1994). "Translating safety, efficacy and compliance into economic value for controlled release dosage forms." Pharmacoeconomics 5(6): 482-504.

Guy, R. H. and J. Hadgraft (1988). "Physicochemical aspects of percutaneous penetration and its enhancement." Pharm Res 5(12): 753-8.

Guy, R. H., J. Hadgraft, et al. (1987). "Transdermal drug delivery and cutaneous metabolism." Xenobiotica 17(3): 325-43.

Karande, P., A. Jain, et al. (2004). "Discovery of transdermal penetration enhancers by high-throughput screening." Nat Biotechnol 22(2): 192-7.

Kobayashi, D., T. Matsuzawa, et al. (1993). "Feasibility of use of several cardiovascular agents in transdermal therapeutic systems with l-menthol-ethanol system on hairless rat and human skin." Biol Pharm Bull 16(3): 254-8.

Lee, C. K., T. Uchida, et al. (1994). "Relationship between lipophilicity and skin permeability of various drugs from an ethanol/water/lauric acid system." Biol Pharm Bull 17(10): 1421-4.

Lee, P. J., R. Langer, et al. (2005). "Role of n-methyl pyrrolidone in the enhancement of aqueous phase transdermal transport." J Pharm Sci 94(4): 912-7.

Mitragotri, S., D. A. Edwards, et al. (1995). "A mechanistic study of ultrasonically-enhanced transdermal drug delivery." J Pharm Sci 84(6): 697-706.

Pino, C., Scherer, MA, Shastri VP (2008). "Percutaneous Delivery of Water Soluble Anesthetics From Triterpene Saponin Glycoside Formulations." J Pharm Sci (in press).

Prausnitz, M. R., V. G. Bose, et al. (1993). "Electroporation of mammalian skin: a mechanism to enhance transdermal drug delivery." Proc Natl Acad Sci U S A 90(22): 10504-8.

Prisant, L. M., B. Bottini, et al. (1992). "Novel drug-delivery systems for hypertension." Am J Med 93(2A): 45S-55S.

Sclar, D. A., T. L. Skaer, et al. (1991). "Utility of a transdermal delivery system for antihypertensive therapy. Part 1." Am J Med 91(1A): 50S-56S.

Singh, J. and M. S. Roberts (1989). "Transdermal delivery of drugs by iontophoresis: a review." Drug Des Deliv 4(1): 1-12.

(Nano)materials in Drug Delivery

Mater. Res. Soc. Symp. Proc. Vol. 1140 © 2009 Materials Research Society   1140-HH04-05

Nanostructured Ceramic Coatings for Drug Delivery

Karin Dittmar[1], Arnaud Tourvieille de Labrouhe[2], Laurent-Dominique Piveteau[2] and Heinrich Hofmann[1]
[1]Laboratory of Powder Technology, Ecole Polytechnique Fédérale de Lausanne, Lausanne 1015, Switzerland
[2]Debiotech SA, Lausanne 1004, Switzerland

ABSTRACT

Drug eluting stents (DES) have been successfully implemented in clinical practice in the treatment against coronary artery disease. Compared to the implantation of bare metal stents, restenosis rates decreased significantly after stenting of DES into coronary arteries [1, 2, 3]. This project aims at creating a novel, structured, ceramic drug delivering coating for stent implants. In the following, the preparation of a titanium dioxide (TiO_2) film with macroporous drug reservoirs in a mesoporous bulk ceramic is shown. The film is predominantly composed of anatase. The broad pore size distribution has a median pore width below 100 nm. The open porosity of the mesoporous bulk, which is greater than 50 %, and the embedded macropores were successfully loaded with Paclitaxel (PTX). The quantity of drug loaded reaches values up to 1.2 $\mu g/mm^2$. The continuous release of the agent into water extends over 1 month. The biocompatibility of the non PTX-loaded, nanostructured coating tested with primary bovine endothelial cells shows good results regarding the number of living cells, the cell vitality and morphology.

INTRODUCTION

Many different materials are under current investigation for the improvement of stents, since restenosis rates of 20-30 % after 6 month of implantation of bare metal stents are recorded [1, 2, 3]. The variety of support materials ranges from non degradable 316L stainless steel or polyurethane polymers to degradable magnesium and polylactic acid. A diversity of coating materials is also in the focus of interest: inorganic carbon, non degradable polymer films, degradable polyglycolic acid, as well as biological components among many others [4]. A metal support and a non-degradable polymer coating storing a drug are often combined to create DES. Once implanted, the drug diffuses out of the polymer with a defined rate. Some commercially available stents with this concept are the Cypher® Stent from Cordis, Taxus® from Boston Scientific and the Endeavor® from Medtronic Vasular [1, 2]. The incorporated pharmacological agents are either Sirolimus, Paclitaxel or derivates thereof. Reports of the implications of implanted DES show a reduction of the restenosis rates to less than 10 % [1, 2, 3]. Nevertheless late studies emphasized problems with the application of these polymer based DES, like a contribution to stent thrombosis as a result of the delayed reendothelialization. Hypersensitivity reactions might be associated to the presence of the polymer and its fragments after delamination, since they occur more than 4 month after the drug release [2]. During this study the feasibility of the application of a well structured coating based on TiO_2 for stent implants is explored. As the native oxide layer of titanium it has already found application in cardiac valves, orthopedic implants and is a well studied biocompatible material [5, 6, 7, 8, 9]. Other challenges during this study are the creation of a prolonged drug release from the coating (> 30 days) to prevent restenosis on a long-term period by inhibiting neoinitmal hyperplasia.

EXPERIMENT

The coating is produced by a multiple-step dip coating procedure on stainless steel or silicon supports. White anatase nanoparticles (d_{BET} = 21.50 nm; TechPowder SA, Switzerland) from a stable colloidal suspension and polymer template microspheres are deposited. The films are heat treated under continuous gas flow and varying atmosphere of air and argon. The maximum temperature applied during the cycle with a gradient temperature profile is 730 °C and the dwell time is 80 min.

The coating is characterized by scanning electron microscopy (XL-30SFEG, Philips) and X-ray diffraction (Powder X' Pert Pro instrument, PANalytical) combined with quantitative phase analysis using the Rietveld refinement method [10]. The specific surface area is measured by nitrogen adsorption (BET model, Gemini 2375, Micrometrics) and the coating's chemical composition is determined by X-ray photoelectron spectroscopy (XPS) on a Kratos AXIS ULTRA instrument. Porosity data and the pore size distributions of the films scratched off the supports after sintering are obtained by mercury intrusion porosimetry (MIP) on Pascal 140, and Pascal 440 instruments from Thermo Electron Corporation. Small angle neutron scattering porosimetry (SANS) with the SANSII beam line at Paul Scherrer Institute (PSI) in Villigen, Switzerland, reveals information about the pore size distribution when the films are not detached from the supports. The average roughness R_a is determined by AFM topography (CPII, Veeco).

The loading of PTX into the porous film is performed with a low pressure solvent evaporation technique using a PTX / solvent solution. Quantification of the drug amount in the coating is assessed by first eluting PTX into acetonitrile and subsequently analysing the solution by high performance liquid chromatography (HPLC) with UV detection at 227 nm (Alliance system from Waters, Photodiode Array Detector 996, Inertsil ODS-3 column). For reporting the continuously released quantity of PTX into ultra pure water as a function of time, the elution medium is totally replaced at defined time points and analysed by HPLC. Samples investigated vary in their structure as well as in their load method. The purely mesoporous coating is loaded using a first method (method 1) while the film containing macroporous drug reservoirs is filled with a second method (method 2).

A biocompatibility test of the TiO$_2$ coating non loaded with PTX is conducted using primary bovine endothelial cells. Cultivation is carried out in DMEM medium supplemented with 10 % fetal bovine serum and antibiotic/antimycotic solution PSA (Invitrogen). After 8 days of incubation, the number of living cells is evaluated. The MTT test (Thiazolyl Blue Tetrazolium Bromide, Sigma) is used to estimate the mitochondrial activity of the cells. The resazurin assay (Resazurin Sodium Salt, Sigma) is performed to investigate possible damages of the cytoplasmic cell membrane. Further microscopic analyses reveal information about the cell morphology. Samples used are mesoporous coatings, coatings with meso and macropores as well as glass cover slips as the negative reference not producing cytotoxic responses [11].

DISCUSSION

The microstructure of the coating is shown in Figure 1a. One can identify the mesoporous bulk TiO$_2$ enclosing ellipsoidal macropores, which are created by the burning out of the polymer templates. The thin film is characterized by an anatase phase content of 79 ± 7 %wt, the rest being rutile. These findings are coherent with thermodynamic studies of the phase transformation of nanosized TiO_2 particles performed by Ranade *et al.* and Kim *et al.* [12, 13]. The phase transformation from anatase to rutile takes place between 700 and 800 °C sinter temperature and depends on the particle size [12, 13]. From the specific surface area of the film, $S_{BET} = 16.68 \pm 1.07$ m^2/g, a mean equivalent diameter of $d_{BET} = 91.35 \pm 5.52$ nm is deduced. For this particle size the most stable phase is anatase. The presence of the minor amount of the polymorph rutile is associated to the presence of bigger grains [12], since the coating exhibits a grain size distribution.

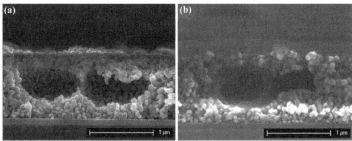

Figure 1. Microstructure of the TiO_2 coating with (a) empty macropores in the mesoporous bulk and (b) macropores as drug reservoirs are filled with PTX.

The surface topography and chemical composition of implant materials have an influence on their integration into biological environments [6, 14]. The mean roughness, R_a, obtained by AFM topography equals 22.7 ± 2.7 nm. Surfaces with a roughness smaller than 50 nm have been shown to increase hemocompatibility by triggering protein adsorption rather than platelet attachment [6, 9]. This might potentially help to prevent thrombus formation after stenting [9], but needs to be investigated in future work. The chemical composition of the coating consists of stoichiometric TiO_2 as confirmed by XPS measurements, Figure 2. TiO_2 being negatively charged when implanted (isoelectric point between pH = 5 to 6, [15]) retards the attachment of blood components with negative charge and thus should also help against thrombus formation as suggested by Liu *et al.* [9].

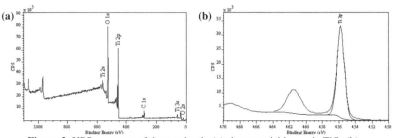

Figure 2. XPS spectrum of the coating in (a) shows stoichiometric TiO_2 (b).

MIP and SANS porosimetry reveal the pore size distribution of the mesoporous bulk. The median equivalent spherical diameter from the volume distribution (D_{v50}) of the SANS measurements is 45 nm, whereas the distribution extends from a few nm to 100 nm. MIP reveals also a broad distribution from 6 nm up to 200 nm with a $D_{v50} = 76$ nm. The differences of the obtained pore width are accounted to the different methods used and the models behind them. In SANS, the closed and the open pores in the coating are detected, whereas in MIP only the open pores are recorded. In SANS, a model using spherical geometry of the pores is applied as compared to a model with cylindrical ones in MIP. Additionally, both methods have different detection limits: in SANS it is restricted by the maximum detectable scatter length of 100 nm whereas in MIP it ranges from 3.5 nm to a scale of several micrometers. Nevertheless one can conclude that the detected pore sizes are in the same range. The open porosity of the bulk is > 50 % as determined by MIP.

The high amount of open pores and voids is successfully loaded with PTX, as demonstrated in Figure 1b. Compared to the non loaded structure in Figure 1a, PTX is accumulated close to the edges of the drug reservoirs. The load of the drug varies from as small as 0.06 μg/mm^2 to a maximum load of 1.21 $\pm$ 0.1 μg/mm^2, depending on the coating's structure and load parameters used. This quantity of pharmaceutical incorporated is comparable with the amount of PTX in commercially available stents [2, 4]. Release profiles of PTX from the coating when immersed into ultrapure water are shown in Figure 3. The drug embedded in the purely mesoporous coating by method 1 diffuses continuously out into the media until 100 % are released after 4 weeks. Up to 16 days PTX is liberated consecutively from a coating containing macropores and only 7 % of the incorporated agent is released (test still proceeding). In the future these experimental results will be compared to molecular dynamics simulations, which will help to identify the diffusion limiting factors of the porous coating. Those might be the pore size, their tortuosity, the PTX adsorption and dissolution rate. The simulation method will then serve as a fundament for understanding the transport governing factors in further release studies using more complex, physiological elution media like blood plasma.

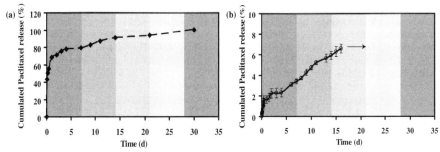

Figure 3. Release curves of PTX from different coating structures into water; (a): mesoporous coating with a load by method 1: 0.18 μg/mm^2; (b): macroporous coating with load by method 2: 0.94 μg/mm^2. The arrow illustrates a test which is still continuing.

In the biocompatibility test a smaller living cell density (LCD) is detected for the samples with TiO$_2$ coating as compared to the reference, Figure 4a. The lowest one with 2000 living cells is recorded for the substrate with meso and macropores. Considering the mitochondrial activity

(Figure 4b) and the cell membrane permeability (results not shown), no significant differences are obtained. Observing the cell morphology by light microscope after staining (photos not shown), one can detect slightly bigger cells on the meso and macroporous samples compared to the blank. This results in a lower density of cells per area of coating. One reason for this behavior is the reaction of the primary endothelial cells to the surface chemistry: borosilicate glass of the reference versus stoichiometric TiO_2 on the coating. Another cause is associated to the impact of the nanotopography of the coating on the cell shape and vitality [14, 15]. A study of Dalby *et al.* with endothelial cells on polymer surfaces of different roughnesses revealed that after 1 week of incubation the cells were more spread on the manufactured rough topographies than on chemically similar flat surfaces. Additionally, in the field of dental and orthopedic medicine it is well demonstrated that nanotopography has a big impact on the osseointegration of implants [14].

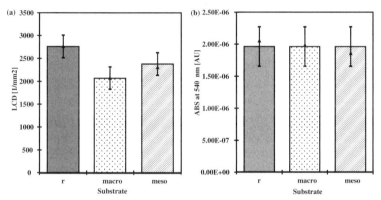

Figure 4. Results of the biocompatibility assay as a function of different substrates: *r*: reference; *macro*: substrate containing macropores; *meso*: purely mesoporous structure. In (a) the living cell density (cells per substrate area) and in (b) their mitochondrial activity are displayed. Results are presented as adjusted means ± SEM after ANOVA analysis of variance (Plan Statistique d'Expériences, J. Lemaitre, EPFL); differences are considered as significant at p<0.05.

CONCLUSIONS

A nanostructured TiO_2 coating, comprising a mesoporous bulk surrounding macroporous drug reservoirs has successfully been created. PTX as a pharmaceutical was accumulated partially in the coating's drug reservoirs and the mesoporous bulk. The maximum load obtained is comparable to the one in commercially available stents. By studying the release of the drug from a purely mesoporous coating over time into water, a continuous release was obtained over 4 weeks. A second test with macropores present in the coating shows a continuous release up to 16 days, where only 7 % of incorporated PTX are liberated (test still proceeding). In the biocompatibility assay, the cell vitality showed no significant difference for the TiO_2 coating as compared to the reference. Differences in the cell morphology are explainable by the different surface chemistries and nanotopographies.

The reaction of the coating to mechanical constraints and especially to the opening of the stent is potentially an issue. The use of thin layers of ceramics is known to be favorable to the mechanical stability. The validity of this approach will be discussed and established in a future publication.

This new coating technology is not limited to the application in stent implants and delivering PTX and/or prohealing agents. Moreover one can think about its usage in orthopedic or dental implants, where the coating might release antibiotics to prevent inflammations besides triggering osseointegration.

ACKNOWLEDGMENTS

The authors would like to thank the LMC of EPFL for XRD and MIP measurements, LMCH at EPFL for XPS measurements and Dr. J. Kohlbrecher at PSI, Villigen for SANS measurements. Research of the project Nanostent (8076.1 LSPP-LS) was supported by the Innovation Promotion Agency, CTI, Switzerland. This project was conducted in collaboration with Debiotech SA who owns exclusive rights on the technology.

REFERENCES

1. S. Venkatraman and F. Boey, "Release profiles in drug-eluting stents: Issues and uncertainties," *Journal of Controlled Release*, vol. 120, Jul. 2007, pp. 149-160.
2. N. Kukreja et al., "The future of drug-eluting stents," *Pharmacological Research*, vol. 57, Mar. 2008, pp. 171-180.
3. http://www.fda.gov/bbs/topics/NEWS/2003/NEW00896.html, October 2008.
4. G. Mani et al., "Coronary stents: A materials perspective," *Biomaterials*, vol. 28, Mar. 2007, pp. 1689-1710.
5. F. Zhang et al., "In vivo investigation of blood compatibility of titanium oxide films," *Journal of Biomedical Materials Research*, vol. 42, 1998, pp. 128-133.
6. D. Buddy, S. Allan, J. Frederick, E. Jack, "Biomaterials Science: An Introduction to Materials in Medicine," Academic Press, 1996.
7. F. Akin et al., "Preparation and analysis of macroporous TiO_2 films on Ti surfaces for bone-tissue implants," *Journal of Biomedical Materials Research*, vol. 57, Dec. 2001, pp. 588-596.
8. N. Huang et al., "Hemocompatibility of titanium oxide films," *Biomaterials*, vol. 24, Jun. 2003, pp. 2177-2187.
9. J. Liu et al., "Sol-gel deposited TiO_2 film on NiTi surgical alloy for biocompatibility improvement," *Thin Solid Films*, vol. 429, April 2003, pp. 225-230.
10. H.M. Rietveld, "A profile refinement method for nuclear and magnetic structures," *Journal of Applied Crystallography*, vol. 2, Jun. 1969, pp. 65-71.
11. "Cell Adhesion and Growth on Coated or modified Glass or plastic surfaces," *Bulletin No. 13, Thermo Fisher Scientific (Nunc GmbH & Co. KG)*, http://www.nuncbrand.com/us/frame.aspx?ID=593, November 2008.
12. M. Ranade et al., "Energetics of nanocrystalline TiO2," *Proceedings of the National Academy of Sciences of the Unites States of America*, vol. 99, Apr. 2002, pp. 6476-6481.

13. C. Kim et al., "Synthesis and particle size effect on the phase transformation of nanocrystalline TiO_2," *Materials Science and Engineering: C*, vol. 27, Sep. 2007, pp. 1343-1346.
14. G. Mendonça et al., "Advancing dental implant surface technology - From micron- to nanotopography," *Biomaterials*, vol. 29, Oct. 2008, pp. 3822-3835.
15. M.J. Dalby et al., "In vitro reaction of endothelial cells to polymer demixed nanotopography," *Biomaterials*, vol. 23, 2002, pp. 2945-2954.

Mater. Res. Soc. Symp. Proc. Vol. 1140 © 2009 Materials Research Society

Encapsulation of Biomolecular Therapeutics into Degradable Polymer Capsules

Siow-Feng Chong, Alisa L. Becker, Alexander N. Zelikin, Frank Caruso
Centre for Nanoscience and Nanotechnology, Department of Chemical and Biomolecular
Engineering, The University of Melbourne, Parkville, Victoria, Australia 3010

ABSTRACT

We report on a novel technique to achieve encapsulation of nucleic acids and oligopeptides, two important classes of biomolecular therapeutics, into degradable polymer microcapsules for diverse biomedical applications. The capsules were obtained from poly(methacrylic acid) chains cross-linked via disulfide bonds and effectively confined the cargo using a combination of steric and electrostatic factors to hinder the diffusion of like-charged macromolecules. While DNA was successfully encapsulated in its pristine form, confinement of the oligopeptides required their conjugation to a carrier polymer. In both cases, assembly protocol allowed control over the capsules loading and was proven to be non-destructive to the encapsulated cargo.

INTRODUCTION

Therapeutics based on biomolecules, such as nucleic acids and peptides, hold promise for the treatment of many diseases. However, these molecules are rapidly degraded in the blood stream, and when administered in their naked form most never reach the site of action. To realize the full potential of these treatments, they must be protected from degradation while in the blood stream. Encapsulation within microscopic delivery vehicles provides a physical barrier to restrict nuclease or protease access. This involves carrier systems such as microparticles [1], liposomes [2], lipid emulsion [3], and polymers [4]. Hollow polymeric microcapsules are a more recent development and have already attracted interest for biomedical applications. This is largely due to the facile manipulation of various properties, including biocompatibility / toxicity, size, structure, and functionality [5,6]. Hollow microcapsules can be created using the layer-by-layer (LbL) assembly technique [7] for thin film formation on a colloidal support in the size range from tens of nanometers to several microns. When the colloidal core is dissolved away, a hollow shell remains.

For successful biomedical applications, the design of a novel colloidal drug carrier must address both the properties of the carrier, such as its biocompatibility and degradation, and the techniques to achieve efficient, reliable and controlled drug loading. In this work we outline the preparation of colloidally stable, monodisperse, biodegradable capsules filled with biomolecules. Capsules formed from disulfide cross-linked poly(methacrylic acid) (PMA_{SH}) can be degraded in a reducing environment, such as that found in the cytosol. We demonstrate the encapsulation of two types of biomolecules with great promise for novel therapeutics: oligopeptides, such as those used as potential HIV vaccines, and DNA for use in antisense or gene therapy. Evidence for the encapsulation of both DNA and oligopeptides within PMA capsules is presented, including data supporting control over the amount encapsulated, and integrity of the encapsulated cargo. Taken together, the developed techniques present a novel platform for diverse biomedical applications, from encapsulated catalysis and sensing to drug delivery.

EXPERIMENT

Materials

The preparation and synthesis of PMA$_{SH}$ [8], 800 bp dsDNA [9], and SiO$_2^+$ particles [9] were all performed as reported previously.

Synthesis of PMA-KP9 Conjugates

A solution of PMA$_{SH}$ with 5 mol % thiol groups was incubated with 1 M concentration of sodium borohydride for 2 h. Concentrated HCl was added to neutralize excess borohydride and mixture was supplemented with phosphate buffer (0.1 M; pH 7.5) and pH was adjusted to pH 8. To this solution, excess Ellman's reagent was added and incubated overnight. The activated PMA$_{SH}$ polymer was purified via NAP-25 columns and freeze-dried.

Activated PMA$_{SH}$ was dissolved in Tris-EDTA buffer (10 mM; pH 7.5). To this solution, fluorescently-labeled KP9 in MilliQ was added and left to proceed overnight. The reaction mixture was purified through NAP-25 columns twice and PMA-KP9 conjugates were recovered by freeze-drying.

Preparation of Oligopeptide-Containing Capsules

1 μm SiO$_2^+$ particles in sodium acetate buffer (20 mM; pH 4) were incubated with varying concentrations of PMA-KP9 conjugates corresponding to 100%, 65%, 35%, and 0% of to surface coverage. The conjugates-coated particles were incubated with PVPON for 15 min. After washing 3 times in sodium acetate buffer (20 mM; pH 4), PMA$_{SH}$ was adsorbed for 15 min. Polymers were added sequentially until 8 layers had been deposited. To stabilise the capsule membrane, the layered particles were exposed to Chloramine T in 2-morpholinoethanesulfonic acid (MES) buffer (20 mM; pH 6) for 1 min. Finally, the particles were suspended in HF/NH$_4$F (2:8 M; pH 5) to remove the template core and the capsules were washed four times.

Preparation of DNA-Containing Capsules

1 μm SiO$_2^+$ particles in sodium acetate buffer (10 mM; pH 4) were incubated DNA corresponding to 50% of the surface saturation value. The DNA-coated particles were incubated with PMA$_{SH}$ for 15 min. After washing 3 times in sodium acetate buffer (10 mM; pH 4), PVPON was adsorbed for 15 min. Polymers were added sequentially until 10 layers had been deposited. The core-shell particles were treated with Chloramine T in MES buffer (10 mM; pH 6) for 2 min in order to cross-link the multilayer. The particles were suspended in sodium acetate buffer (10 mM; pH 4) to which HF/NH$_4$F (2:8 M; pH 5) was added. This step removed the template core, leaving PMA capsules loaded with DNA.

Gel electrophoresis of the Encapsulated DNA

DNA-containing capsules were prepared as described above using 1 μm diameter SiO$_2^+$ particles and non-thiolated PMA and PVPON as capsule constituting polymers. Non-thiolated polymers were used to facilitate capsule degradation without use of reducing agents, which inhibit PCR. For analysis, the capsules were suspended in the PCR buffer, which resulted in immediate deconstruction of the capsules and release of the encapsulated DNA. The released

DNA, along with empty capsules and stock DNA were analyzed on a 1% agarose gel in TAE buffer with GeneRuler 1kb DNA ladder, and stained with ethidium bromide.

RESULTS AND DISCUSSION

To prepare capsules suitable for the encapsulation of biomolecules, thiol-functionalized poly(methacrylic acid) (PMA_{SH}) and poly(vinylpyrrolidone) (PVPON) were deposited alternately on a sacrificial colloidal template using hydrogen bonding for thin film build-up. After conversion of thiol groups into disulfide linkages and core removal, hollow capsules were obtained. Using disulfide cross-links between PMA layers gave rise to redox-active deconstructable capsules. The reducing environment within a cell can reduce the disulfides into thiols, which then results in the deconstruction of capsules and the release of their cargo. Properties such as stability, swellability, permeability, pH sensitivity allow pre- and post-encapsulation [10] of therapeutics within PMA capsules. The former method display efficient encapsulation and allows great control over the quantities loaded. Here, we compare and contrast the encapsulation of DNA and oligopeptides, two biomolecules with therapeutic potential that are not effectively delivered by conventional means.

Previously, PMA capsules have been used to encapsulate proteins [11]. The encapsulation was achieved by the physical barrier created by the capsule membrane against the entrapped cargo. Another potential factor that governs the encapsulation within a negatively charged PMA capsule is through the repulsion force generated by the capsule wall and cargo. This is an advantage for the encapsulation of DNA as the backbone of DNA is negatively charged. To encapsulate DNA, it is first adsorbed onto amine-functionalized, positively charged particles (SiO_2^+) via electrostatic interactions, followed by the alternative assembly of PMA_{SH} and PVPON at low pH conditions to form a hydrogen-bonded, pH-sensitive thin film. The thiol groups within the PMA_{SH} film were then cross-linked to form disulfide linkages (see Figure 1), followed by template removal to yield polymer capsules. This method has been used to encapsulate DNA from 20-mer to 3 kbp [9, 12].

In contrast, the oligopeptide KP9 is too small to be confined within the semi-permeable PMA capsule. To encapsulate the oligopeptides, they were first modified with a cysteine residue and conjugated to PMA_{SH} through disulfide linkages. This increased the effective molecular weight of the oligopeptide, and provided the same charge as the capsule. After conjugation, the PMA-conjugated oligopeptides (PMA-KP9) were deposited onto positively charged particles, and encapsulated using the same procedure as the DNA-loaded capsules (see Figure 1).

Both materials were successfully encapsulated using the pre-loading technique of adsorbing the biomolecules onto the template prior to thin film formation. When the capsules were transferred into pH 7 solution, the PMA became ionized and the capsules swelled. As the DNA is negatively charged, and the oligopeptides are conjugated to a negatively charged carrier, the repulsion from the capsule wall expels the cargo to the interior of the capsule (see Figure 2). Although they differ in their pristine forms (e.g. size, ionization, structure, hydrophilicity etc.), the PMA-KP9 and DNA were both retained in the microcapsule by steric hindrance and charge repulsion of the polymer film. Despite these similarities, there are significant differences in the formation and properties of the capsules.

61

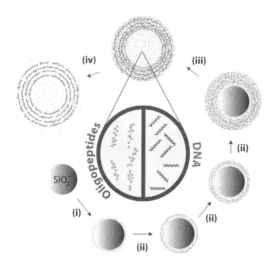

Figure 1. Encapsulation of biomolecules into degradable polymeric capsule: (i) adsorption of biomolecules onto amine-functionalized silica particle; (ii) assembly of a thin polymer film prepared via the alternating deposition of PVPON and PMA$_{SH}$; (iii) oxidation of PMA$_{SH}$ thiol groups into bridging disulfide linkages and removal of core particle, result in stable polymer capsule; (iv) capsule destruction, releasing biomolecules in a reducing environment

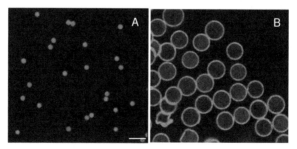

Figure 2. Confocal laser scanning microscopy (CLSM) images of (A) 1 μm 4-bilayer PMA capsules filled with PMA-KP9 conjugates, KP9 is labeled green; scale bar is 5 μm and (B) 3 μm 4-bilayer PMA capsules filled with 800 bp dsDNA, capsule wall is labeled green, DNA is labeled red.

For example, the amount of DNA on the surface of the particles cannot exceed 50% saturation coverage, as this decreases subsequent LbL and leads to unstable capsules [9]. In contrast, this is not observed for the PMA-KP9-coated particles. As the oligopeptides were covalently anchored onto the PMA$_{SH}$ polymer, the adsorption of PMA-KP9 serves as the first layer of the capsule membrane. This does not affect the sequential adsorption of PVPON and PMA$_{SH}$. In addition, the quantity of oligopeptides conjugated to the PMA can be varied; and the conjugates can be incorporated in the thin film several times. Thus, there is no apparent limit on the amount of oligopeptides that can be encapsulated. This provides a greater control over the quantity of drug encapsulated and hence the pharmacokinetics properties of therapeutic in the body. Indeed, simply by varying the concentration of PMA-KP9 incubated with the particles, the amount of drug can be controlled. Unlike the encapsulation of DNA where the surface coverage was limited to partial saturation, the adsorption of conjugates can be 100% without affecting the deposition of the next PVPON layer.

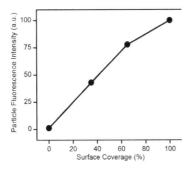

Figure 3. Fluorescence of 1 μm capsule with varied loadings of the PMA-KP9 conjugates.

The disulfide-stabilized PMA capsules have been shown to disintegrate within 4 hours of incubation in the presence of 5 mM reduced glutathione [11]. This condition mimics the reducing environment of the cell, suggesting the confined DNA and oligopeptides within such capsules can be released intracellularly upon the internalization of capsules by the live cells.

Due to the different encapsulation methods, the release of the cargo in a reducing environment is likely to be different. For instance, the small oligopeptide is expected to permeate the capsule wall and be released once the disulfide bond holding it to the carrier polymer is broken. Thus, oligopeptides release can occur before the capsule has degraded. In contrast, DNA release requires a hole to be formed in the microcapsule wall and therefore may require more advanced degradation of the capsule prior to cargo release. The release of oligopeptides and DNA from capsules in reducing conditions is currently being investigated.

For biomedical purposes, the encapsulated cargo should maintain its integrity and functionality. To investigate this, DNA that was encapsulated and released was analyzed by gel electrophoresis (see Figure 3). The released DNA has the same electrophoretic mobility (i.e. size and charge) as the pristine stock DNA. We have also shown that the released DNA can be used in the polymerase chain reaction (PCR) [9]. PCR requires protein-DNA recognition and DNA hybridization for its success. These are good indicators that the DNA remains intact and functional. On the other hand, the functionality of the oligopeptide has recently been proven in whole blood [13]. Capsules were internalized by dendritic cells and the peptide was released. The functional oligopeptides cause the presentation of MHC class I molecules to T cells, resulting in the stimulation of T cells to produce cytokines. The production of cytokines was used as the indicator of oligopeptides release and activity, showing the chemical modification does not affect the functionality of the encapsulated oligopeptides.

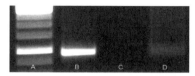

Figure 4. Gel electrophoresis image of A) DNA ladder B) stock DNA C) Empty 1 μm PMA capsules D) DNA released from 1 μm PMA_{SH} capsules.

CONCLUSIONS

DNA and oligopeptides were successfully encapsulated within PMA capsules. There were some similarities between the encapsulations of DNA and oligopeptides: both biomolecules were encapsulated using a pre-loading technique, steric hindrance and ionic repulsion were used to trap the cargo, which remained intact and functional. The differences were that the DNA was encapsulated in pristine form, whereas the oligopeptides were first conjugated to a carrier polymer. This provided more control over the amount of oligopeptides to be encapsulated, as it did not affect subsequent LbL. The PMA capsules have provided various encapsulation possibilities for biomolecular therapeutics, which allow their use in biomedical applications.

ACKNOWLEDGMENTS

This work was supported by the Australian Research Council via the Australian Postdoctoral Fellowship (ANZ), Discovery Project, and Federation Fellowship (FC) schemes. The Particulate Fluids Processing Centre is acknowledged for infrastructure support.

REFERENCES

1.   D. A. Edwards, J. Hanes, G. Caponetti, J. Hrkach, A. Ben-Jebria, M. L. Eskew, J. Mintzes, D. Deaver, N. Lotan, R. Langer, *Science* **276**, 1868 (1997).
2.   V. P. Torchilin, *Nat. Rev. Drug Disc.* **4**, 145 (2005).
3.   R. N. Gursoy, S. Benita, *Biomed. Pharmacother.* **58**, 173 (2004).
4.   R. Duncan, *Nat. Rev. Drug Disc.* **2**, 347 (2003).
5.   D. B. Shenoy, A. A. Antipov, G. B. Sukhorukov, H. Möhwald, *Biomacromolecules* **4**, 265 (2003).
6.   G. Schneider, G. Decher, *Langmuir* **24**, 1778 (2008).
7.   E. Donath, G. B. Sukhorukov, F. Caruso, S. A. Davis, H. Möhwald, *Angew. Chem. Int. Ed.* **37**, 2202 (1998).
8.   A. N. Zelikin, J. F. Quinn, F. Caruso, *Biomacromolecules* **7**, 27 (2006).
9.   A. N. Zelikin *et al.*, *ACS Nano* **1**, 63 (2007).
10.   A. P. R. Johnston, C. Cortez, A. S. Angelatos, F. Caruso, *Curr. Opin. Colloid Interface Sci.* **11**, 203 (2006).
11.   A. N. Zelikin, Q. Li, F. Caruso, *Chem. Mater.* **20**, 2655 (2008).
12.   A. N. Zelikin, Q. Li, F. Caruso, *Angew. Chem., Int. Ed.* **45**, 7743 (2006).
13.   R. De Rose, A. N. Zelikin, A. P. R. Johnston, A. Sexton, S.-F. Chong, C. Cortez, W. Mulholland, F. Caruso, S. J. Kent, *Adv. Mater.* **20**, 4698 (2008).

Mater. Res. Soc. Symp. Proc. Vol. 1140 © 2009 Materials Research Society   1140-HH12-05

Surface Engineering for *in vivo* Targeting Delivery of siRNA for Efficient Cancer Therapy

O. Taratula[1,2], P. Kirkpatrick[1], R. Savla[1,2], I. Pandya[1], T. Minko[2,3], H. He[1,3]

[1] Department of Chemistry, Rutgers University, 73 Warren Street, Newark, NJ 07102, U.S.A.
[2] Department of Pharmaceutics, Ernest Mario School of Pharmacy, Rutgers, The State University of New Jersey, Piscataway, NJ 08854, U.S.A.
[3] Cancer Institute of New Jersey, 195 Little Albany Street, New Brunswick, NJ 08903, U.S.A.

ABSTRACT

The main obstacle in siRNA therapy is delivering RNA to the cytoplasm where it can guide sequence-specific mRNA degradation. Attempts to develop effective nonviral vectors for *in vivo* delivery of nucleic acids through a systemic route have been hampered by difficulties in combining high extracellular stability with availability of the nucleic acids following entry into cells. Other challenges with non-viral gene delivery include limitations in target-cell specificity. Here we demonstrate that the caging of siRNA-PPI nanoparticles with a dithiol, containing crosslinkers, provides lateral stabilization, preventing unfavorable dissociation of the nanopartices before entering the cytoplasm of target cells. Further PEGylation of the caged nanoparticles stabilizes them against aggregation induced by salts and proteins in the serum. Due to the reductive environment in the cytoplasm, the disulfide bonds can be reduced, and the lateral crosslinks can be removed, releasing the siRNA for expression. Furthermore, PEG coating can effectively eliminate nonspecific delivery, increasing targeting delivery efficiency after specific targeting groups are attached to its distal ends. Our results demonstrate that the approach successfully stabilized the siRNA nanoparticles against aggregation and unfavorable dissociation in human serum, and it also leads to targeted delivery of the siRNA nanoparticles into specific cancer cells. Most importantly, the internalized siRNAs can be released into the cytoplasm and used to efficiently silence their targeted mRNA.

INTRODUCTION

There is an increasing interest for developing therapies based on RNA interference (RNAi). However, just like the other gene therapy strategies, the main obstacle to the success of siRNA therapeutics is in delivering it across the cell membrane to the cytoplasm, where it can enter the RNAi pathway [1]. Viruses have evolved functions to accomplish this; however, the immune response elicited by viral proteins has posed a major challenge to this approach. Therefore, there is much interest in developing nonviral gene delivery vehicles. Attempts to develop effective nonviral vectors for *in vivo* delivery of nucleic acids through a systemic route are hampered by difficulties of combining of high extracellular stability with readily available nucleic acids following entry into cells. Extracellular stability is essential as the delivery system should be capable of withstanding the aggressive biological environment en-route to the target site, while availability of the nucleic acids permits efficient therapeutic effects within the cells. In order to overcome the existence obstacles for *in vivo* gene delivery, we started with nanometer size (~100 nm) nanoparticles, formed by packaging of siRNAs with generation 5 poly(propyleneimine) dendrimer (PPI G5). The improvement of the extracellular stabilization of the resulting nanoparticles was achieved by step-by-step modification of their outer surface with a dithiol containing crosslinking agent, dimethyl-3-3'-dithiobispropionimidate-HCl (DTBP) [2], and heterobifunctional poly(ethylene glycol) (PEG), α-maleimide-ω-N-hydroxysuccinimide ester poly(ethylene glycol) (MAL-PEG$_{5000}$-NHS) [3]. Finally, the stabilized nanoparticles were successfully equipped with a targeting capacity by covalently attaching a peptide to the distal end of PEG to direct the siRNA nanoparticles only to the targeted cancer cells.

EXPERIMENTAL DETAILS

The condensed siRNA complexes were prepared at a desired 2.4 amine/phosphate (N/P) ratio either in DI water or 10 mM Hepes buffer (pH 7.2) by adding a stock solution of PPI G5 dendrimer (typically, 500 μM) into the siRNA solution mentioned above. The samples were vortexed briefly, and the solutions were then incubated at room temperature for 30 minutes to ensure complex formation.

In order to crosslink individual complexes, DTBP (2.5 mg/mL) dissolved in HEPES buffer was added to prepared siRNA/PPI G5 complexes solution at various concentrations depending on the desired crosslinking ratio. For example, DTBP: NH_2 = 3.5 indicates siRNA/PPI G5 complex crosslinked with DTBP using molar ratio of 3.5 between DTBP and total amino groups of PPI G5 available after the condensation reaction. After 3 hrs of crosslinking reaction, maleimide-PEG-NHS (35mg/mL) was added to the solution and the NHS groups on the distal end of the PEG reacted with amine groups on the periphery of siRNA/PPI G5 complex for 1hr at room temperature. Next, 12.5 mg/mL of LHRH peptide was added and left overnight to covalently conjugate the peptide on the distal end of the PEG layer on the siRNA/PPI G5 complex through the maleimide groups on the PEG and the thiol groups in the peptide. siRNA/PPI G5 modified complexes were then purified by dialysis (MW cut-off 10 kDa) against water for 1 day and used for further studies.

RESULT AND DISCUSSION

Preparation and Stabilization of siRNA nanoparticles

It has been recognized that a prerequisite for the facile transport of siRNA through the cell membrane is their complexation to nanoparticles ~200 nm size [4]. The nanoparticles can protect the siRNA by sterically blocking the access of nucleolytic enzymes. Our AFM studies demonstrate that PPI G5 dendrimers were able to effectively package siRNA into discrete nanoparticles (Fig. 1A). The nanoparticles appear to be spherical after 30 minutes of complexation with an average diameter of 150.1 ± 22.2 nm and a height of 5.4 ± 0.9 nm from AFM measurements. Nevertheless, the formulated nanoparticles did not remain stable over a long period of time under storage conditions. According to the AFM studies, after 48 hrs in PBS buffer, the nanoparticles sufficiently increased in average size from 150.1 ± 22.2 nm to 612.7 ± 451.1 nm (Fig. 1A and 1C). The evidence of siRNA nanoparticles aggregating under experimental conditions can be problematic for their *in vivo* application, since aggregation can lead to rapid clearance of the gene delivery complexes by phagocytic cells and the reticuloendothelial system [5].

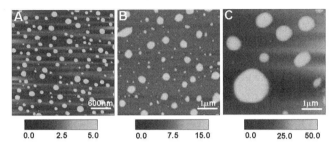

Figure 1. AFM images of siRNA nanoparticles packaged by PPI G5 dendrimer (A) after 30 min of condensation. (B) siRNA nanoparticles modified with DTBP, PEG and LHRH and (C) nonmodified siRNA nanoparticles were stored for 48 hrs at room temperature.

In order to provide lateral stabilization and inhibit the possible replacement of siRNAs by serum polyanions, the formulated siRNA nanoparticles were modified with DTBP. From TNBSA assays we estimated that 26.5% of free amine groups on the siRNA nanoparticles were crosslinked. To prevent the siRNA nanoparticles from aggregating and extend their circulation time in blood streams, we further modified the caged siRNA nanoparticles with a PEG. The stability of the siRNA nanoparticles with covalent stabilization by DTBP and the PEG layer was first evaluated by AFM. The obtained data reveals that in comparison to non-modified particles, the modified ones became slightly larger (256.8 ± 123.8 nm). However, after 48 hours in PBS buffer, the size of the particles stayed almost the same with a relatively narrow distribution, demonstrating they are largely stabilized by the two-layer modification. The increase in the size of the prepared nanoparticles could be attributed to their modification by PEG [6].

The stability of the siRNA nanoparticles with or without covalent stabilization was further studied by the ability of the siRNA nanoparticles against polyanion disruption. The experiments were performed by measuring the ability of a polyanion, polymethacrylic acid (PMAA) to restore siRNA access to ethidium bromide (EtBr). The EtBr fluorescence was dramatically increased upon its incubation with free siRNA (from pink curve to red curve in Figure 2A). Compaction of the siRNA with PPI G5 dendrimers causes the EtBr fluorescence to decrease to the level of free dye (Fig. 2A, black curve). Crosslinking the siRNA nanoparticles with DTBP did not introduce any significant change on the fluorescence, indicating that the siRNA nanoparticles remain compacted during the crosslinking process. PMAA was added to the siRNA nanoparticles solution. As the concentration of PMAA was increased progressively, higher fluorescence readings were observed, which demonstrates, that siRNA release from nonmodified complexes. For the siRNA nanoparticles without DTBP and PEG layer, 5 μM PMAA was able to release 85 % of the siRNA for ethidium bromide binding (Fig. 2B, black curve). For the siRNA nanoparticles caged with DTBP, 5 μM PMAA released only 14% of siRNA. Addition of PMAA up to 100 μM did not cause further release of siRNA, indicating that siRNA nanoparticles caged by DTBP are stable against PMAA disruption (Figure 2B, green curve). When 25 mM glutathione was introduced to promote reduction of the intramolecular disulfide bond in DTBP, the stabilization of siRNA reversed. For the PEG- protected siRNA nanoparticles, with PMAA concentration increase, the fluorescence gradually reached 70% of that observed for the free siRNA, suggesting that PEG layer alone could not protect siRNA nanoparticles from the polyanions disruption (Figure 2B, red curve). With the combination of the

DTBP crosslinking and PEGylation on the siRNA nanoparticles, addition of 5 μM PMAA released 10% of siRNAs, no further release was observed at higher PMAA concentrations tested (blue curve in Figure 2B). Addition of 25 mM glutathione into the solution, led to 80% of siRNA being released from siRNA nanoparticles (light blue curve in Figure 2A). The results demonstrated the efficiency of a layer-by-layer modification approach to prevent the disruption of siRNA complexes in the presence of competing polyanions, which are widely abundant in the extracellular environment. Moreover, the availability of the disulfide bond in the structure of the crosslinking agent provides a triggered release of siRNA in the cytoplasm of the targeted cell.

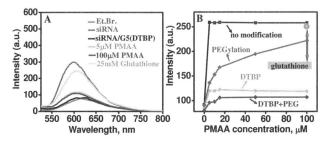

Figure 2. (A) Evolution of EtBr fluorescence spectra of free siRNA, siRNA/PPI G5 nanoparticles and crosslinked with DTBP in the presence of different concentration of PMAA and 25 mM glutathione. (B) EtBr fluorescence from interactions with siRNA nanoparticles without and with DTBP crosslinker, with PEGylation, and with combination of DTBP and PEGylation in the presence of PMAA and 25 mM glutathione.

PEGylation prevents aggregation and nonspecific delivery of siRNA nanoparticle to cells

To further demonstrate if the PEG layer can prevent siRNA nanoparticles from aggregation in extracellular environments, we have compared the ability of the siRNA nanoparticles with different modification to undergo cellular uptake in the A549 cancer cell line. Nonmodified siRNA nanoparticles show serious aggregation in the cell medium (Fig. 3A). In case of the modified nanoparticles, one could barely see any aggregation of the siRNA nanoparticles (Fig. 3B, C), indicating the PEG layer can remarkably prevent aggregation induced by physiological salts and serum proteins. Furthermore, the stabilized siRNA nanoparticles at lower modification ratio (DTBP: NH_2 =3.2:1 in the conjugation solution) dramatically decreased in internalization by the cells, compared to the one with higher ratio (DTBP: NH_2=15.9:1). The higher molecular ratio during the caging process results in decreasing NH_2 groups on the siRNA particles surface for PEG conjugation. Even though the PEG layer prevents the aggregation of the siRNA nanoparticles, while large amount of siRNAs were still internalized by the cells nonspecifically. On the other hand, with lower molecular ratio during the caging process, more NH_2 groups left on the siRNA particles surface for PEG conjugation, so the density of the positive charges on the siRNA nanoparticle surfaces was largely decreased after further PEGylation of the DTBP caged siRNA nanoparticles, allowing more effective delivery of the siRNA nanoparticles to cells.

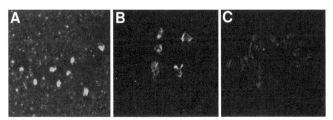

Figure 3. Fluorescence microscopy images of (A) siRNA/PPI G5 complexes (B) siRNA/PPI G5/DTBP/PEG (NH_2 : DTBP = 1: 15.9) and (C) siRNA/PPI G5/DTBP/PEG (NH_2: DTBP = 1: 3.2) after 24 hrs of incubation with A549 cancer cells.

Specific delivery and sequence specific knockdown of the targeted mRNA

Nonspecific delivery of genes toward both cancerous and normal tissues can result in serious side effects, thereby limiting their clinical applications. To deliver the siRNA nanoparticles selectively to the distant tumor metastases after injection into the blood stream, the particles have to be equipped with specific ligands to be delivered specifically into cancer cells. Recently we have successfully used a modified peptide synthetic analog of luteinizing hormone-releasing hormone (LHRH) decapeptide as a targeting moiety to tumors overexpressing LHRH receptors [7]. The use of this peptide enhanced drug accumulation in tumors and its internalization by cancer cells. To specifically deliver the caged siRNA nanoparticles with a PEG layer to cancer cells, we applied a heterobifunctional MAL-PEG-NHS to modify the DTBP caged siRNA nanoparticles. We followed the well-documented coupling procedures to covalently link PEG with amine groups on the caged siRNA nanoparticles (DTBP: NH_2=3.2:1), and then conjugated the LHRH peptides to the distal end of PEG layer through MAL group and the cystein in the LHRH peptides. Fluorescent microscopic studies demonstrated the LHRH-siRNA nanoparticles can be specifically internalized by LHRH positive A549 cells and A2780 cells but not SKOV-3 cells, which expressed a low level of LHRH receptors (Fig. 4).

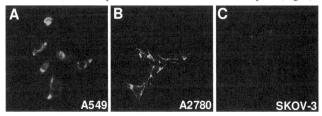

Figure 4. Representative fluorescence microscopic images of cellular uptake of the LHRH-PEG-DTBP-modified FITC-siRNA nanoparticles by LHRH positive, (A) A549 cells, (B) A2780 cells, and LHRH negative (C) SKOV-3 cancer cells.

Theoretically, the formulated siRNA complexes could adhere to the surface of cancer cells and erroneously be visualized on two-dimensional fluorescent images as internalized within cell. To exclude such errors, we analyzed the distribution of the prepared siRNA complexes in different cellular layers from the upper to the lower of the fixed cell using confocal fluorescent microscopy. The z-sections of single cells transfected with the siRNA complexes showed their homogeneous and uniform and the distribution in the different cells layers (data not shown).

A RT-PCR study was also used to study the ability of the peptide conjugated siRNA nanoparticles to silence their target mRNA expression as describe above. The results demonstrated that the siRNAs can effectively knock down their target mRNA, even though different efficiency was shown in different cell lines (Fig. 5). The transfection of LHRH positive A2780 and A549 cancer cell lines with the modified siRNA complexes under the same experimental conditions resulted in 98.0% and 73.3% *BCL2* targeted mRNA suppression, respectively. On the other hand, the silencing effect of the targeted mRNA was not observed in LHRH-negative cancer cells, which are consistent with the fact that the LHRH-negative SKOV-3 cancer cells were not capable to take up the formulated siRNA complexes as represented in Figure 4C.

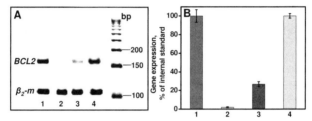

Figure 5. RT-PCR results (A) show the effect of treatment on different cell lines. Lane 1, no treatment; Lane 2, 3, 4 treated by the engineered siRNA nanoparticles on the expression of *BCL2* mRNA, lane 2 for A2870 cells, Lane 3 for A549 cancer cells, Lane 4 for SKOV-3 cancer cells, respectively. (B) represents the quantitative data for *BCL2* mRNA suppression in the above treated cell lines.

CONCLUSIONS

In summary, crosslinking and PEGylation of the formulated nanoparticles make siRNA more stable in the systematic circulation after intravenous delivery. As a result the proposed siRNA delivery system is capable of extracellular stability and effective targeted delivery can be used for *in vivo* systematic delivery of siRNAs for efficient cancer therapy.

ACKNOWLEDGMENTS

The research was supported in part by NIH grants CA100098, CA111766, CA074175 from the National Cancer Institute and Charles & Johanna Bush Biomedical Grants.

REFERENCES

[1] D.M. Dykxhoorn, J. Lieberman, *Annu. Rev. Med.* **23,** 401 (2005).
[2] D. Oupicky, R.C. Carlisle, L.Y. Seymour, *Gene Ther.* **8,** 713 (2001)
[3] V.S. Trubetskoy, A. Loomis, P.M. et al., *Bioconjug. Chem.* **10,** 624 (1999)

[4] V. Vijayanathan, T. Thomas, T.J. Thomas, *Biochemistry* **41,** 14085 (2002)
[5] P.R. Dash, M.L. Read, L.B. Barrett, M.A. Wolfert, L.W. Seymour *Gene Ther.* **6,** 643 (1999)
[6] J.H. van Steenis, E.M. van Maarseveen, F.J. Verbaan et al., *J. Control Release* **87,** 167 (2003)
[7] S.S. Dharap, Y. Wang, P. Chandna, J.J. Khandare, et al., *PNAS* **102** 12962 (2005)

Mater. Res. Soc. Symp. Proc. Vol. 1140 © 2009 Materials Research Society   1140-HH05-17

Antisense oligonucleotides delivery to antigen presenting cells by using schizophyllan

Shinichi Mochizuki[1], Jusaku Minari[1], Kazuo Sakurai[1,2]
[1] The University of Kitakyushu, 1-1, Hibikino, Wakamatsu-ku, Kitakyushu, Fukuoka, 808-0135, Japan
[2] CREST, Japan Science and Technology Agency, 4-1-8, Honcho, Kawaguchi-shi, Saitama, 332-0012, Japan

ABSTRACT

Shizpphyllan (SPG), a kind of β-1,3-glucan, is known to be recognized by dectin-1 on antigen presenting cells. It was also revealed that SPG can form a complex with polynucleotide, and this complex consists of two polysaccharide-strands and one polynuleotide-strand. The complex consisting of SPG and oligonucleotides was prepared and the formation was confirmed by gel permeation chromatography. Furthermore, the complex could inhibit the binding of biotin-labeled SPG to the cells having dectin-1. These data indicate that SPG can be recognized after the complexation with oligonuicleotides (ODNs) and has an ability to deliver ODNs into antigen presenting cells. These finding indicate that β-1,3-glucans are very attractive and useful materials in gene delivery technology.

INTRODUCTION

Recently, the synthetic oligonucleotides, such as antisense and siRNA oligonucleotides, have been shown to have useful applications in the treatment of various diseases, including cancer. Several first generation (phosphorotioate) antisense oligonucleotides are in late phase clinical testing[1], while newer oligonucleotide chemistries are providing antisense molecules with higher binding affinities, greater stability and lower toxicity as clinical candidates[2]. However, *in vivo*, there are a number of obstacles to overcome, such as rapid excretion via kidney, degradation in serum, uptake by phagocytes of the reticuloendothelial system, and inefficient endocytosis by target cells. A variety of supramolecular nanocarriers including liposome, cationic polymer complex and various polymeric nanoparticles have been used to deliver antisense and siRNA oligonucletides, as more fully described in several recent reviews[3]. Complecation of oligonucleotides with various polycations is a popular approach for intracellular delivery, this including use of PEGlyated polycations[4], polyehtyleneimine complexes[5], cationic block co-polymers[6] and dendrimers[7]. However, the large size and/or considerable toxicity[8] of cationic lipid particles and cationic polymers may render them problematic candidates for *in vivo* utilization. Many investigators believe that appropriate delivery carriers could be very useful for oligonucleotide-based therapeutics[9].

We have reported that schizophyllan (SPG, see figure 1 for the chemical structure), a kind of β-1,3-glucan, can form a macromolecular complex with some homo-phosphodiester polynucleotides[10, 11]. SPG, an extracellular polysaccharide produced by the fungus, is consists of β-1,3-glucan and one β-1,6-glycosyl side chain that links to the main chain at every

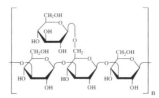

Figure 1. Structure of SPG

three glucose residues. Norisuye et al[12] carefully studied the dilute solution properties of SPG and noted that it dissolves in water as a triple helix (t-SPG) and in dimethylsulfoxide (DMSO) or basic solution (pH > 13) as a single chain (s-SPG). Furthermore, when water is added to an s-SPG/DMSO solution or alkaline solution is neutralized, t-SPG is regenerated from three s-SPG chains through hydrophobic and hydrogen-bonding interactions (known as the renaturation process)[13]. Although the resultant renatured product is not exactly the same as the original rod-like molecule, the local structure was proven to restore the triple helix.

β-glucans are structural components of fungal cell walls with well-characterized immunostimulatory properties and such as have been widely used to study the functions of leukocytes as well as inflammatory processes *in vivo*. Dectin-1 was identified as a major receptor involved in the recognition of these carbohydrates[14, 15]. Dectin-1 is now known to be expressed by many cell types, including macrophages, dendritic cells, monocytes, neutrophils. We have focused on this interaction between β-glucans and dectin-1, and studied SPG as a delivery carrier to the cells having dectin-1.

In this study, we prepared complex consisting of antisense oligonucleotides (AS-ODNs) and SPG, and calculated the rate of complexation by using gel permeation chromatography. We examined whether the complexes are recognized by dectin-1 on the cells.

EXPERIMENT

Materials
Triple strand SPG (M_w; 1.5×10^5) was kindly provided by Taito Co. Ltd., (Kobe, Japan). HEK293 cells transfected with dectin-1 were kindly supplied from Y. Adachi (Tokyo University of Pharmacy and Life Sceince, Tokyo, Japan). 2-aminoethanol, D-biotin N-succinimidyl ester and sodium cyanoborohydride were purchased from Tokyo Kasei Kogyo Co., Ltd. (Tokyo, Japan).

Preparation of biotin-SPG
SPG was partially periodate oxidized from its side-branched glucose residues, and resulting aldehyde groups were conjugated with 2-aminoehtanol by reductive amination with sodium cyanoborohydride. The resultant amino groups were reacted with D-biotin N-succinimidyl ester. The biotinylated SPG (b-SPG) was dialyzed against water and obtained by freeze-drying.

RT-PCR
Total RNA of each cell was prepared with TRIzol reagent. First-strand cDNA was synthesized using the SuperScript® III RNase H⁻ Reverse Transcriptase (Invitrogen, Carlsbad,

CA). The PCR protocol was as follows: 94 °C for 3 min; 25 cycles of 94 °C for 30 s, 60 °C for 30 s and 72 °C for 60 s. The primer sets were as follows: Dectin-1: 5'-TCAGGGAGAAATCCAGAGGA-3' (forward) and 5'-CTTGAAACGAGTTGGGGAAG-3' (reverse); β-actin: 5'-GGCTACAGCTTCACCACCAC-3' (forward) and 5'-AGGGCAGTGATCTCCTTCTG-3' (reverse).

Flow cytometry analysis

Flow cytomery (Epics XL; Beckman Coulter, Fullerton, CA) was performed according to conventional protocols. Cells were examined by using the mAb, 2A11 (Serotec, Oxford, UK) and Alexa488-labeled anti-rat IgG antibody (Invitrogen).

Complexation of SPG/ODNs

Complexation between SPG and AS ODNs (80mer having dA_{60} tail at 3' terminal) was carried out with the established method[11]. The molar ratio ($M_{s\text{-}SPG}/M_{ODN}$) was controlled to 3.0. After the complexation, it was confirmed by gel permeation chromatography (GPC). GPC was carried out using a Shodex (Showa Denko, Tokyo, Japan) DU-H2130 pumping system at the flow rate of 1.0 ml/min with OHpak SB-806 and OHpak SB-802.5 columns (Showa Denko). The aqueous solution containing 50 mM phosphate buffer was used as a mobile phase. The eluate was detected by an ultraviolet (UV) detector (SPD-10A, SHIMADZU, Kyoto, Japan) and a multiangle light scattering (LS) detector (DU-H2130, Wyatt Technology, Santa Barbara, CA).

Analysis of SPG-biotin binding to the cells

Dectin-1 transfectants (3.0×10^5 cells) were suspended in 200 µl of PBS/F/E (5% FBS, 2 mM EDTA). Medium containing 30 µg/ml b-SPG with or without the complex was added to each well and incubated at 37 °C with gentle shaking for 30 min. After the incubation, the cells were washed, stained with streptavidin-Alexa 488 (Invitrogen) and analyzed by flow cytometry.

DISCUSSION

We have prepared HEK293 cells transfected with mouse dectin-1 cDNA (d-HEK293) in order to examine its ability to bind to SPG. The extent of expression of dectin-1 was confirmed by RT-PCR and flow cytometry. Strong expression of dectin-1 mRNA was observed by RT-

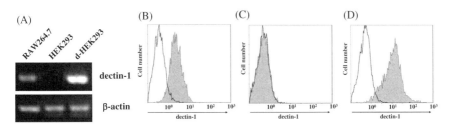

Figure 2. The expression of dectin-1 and β-actin were observed by RT-PCR in RAW264.7, HEK293 and d-HEK293 (A). Dectin-1 expression on RAW 264.7 cell (B) and HEK293 (C) and

d-HEK239 (D) are analyzed by flow cytometry. The open and shaded histograms show the results for unstained and stained with anti-dectin-1 antibody, respectively (B-D).

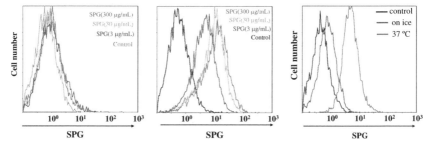

Figure 3. Biotin-SPG can bind to dectin-1 transduced HEK293 cells. Control HEK293 cells (left) and d-HEK293 cells (middle) were treated with 0, 3, 30 and 300 µg/ml of b-SPG. The amount of b-SPG bound to the cells was determined by staining with streptavidin-Alexa 488 conjugate and using a flow cytometer. After RAW264.7 cells were incubated with Alexa 488-labeled SPG (3 µg/ml) at 37 °C or on ice for 30 min, the fluorescence intensity of the cells was examined by using flow cytometer (right).

PCR analysis in the d-HEK 293 cells, but almost not at all in the HEK293 cells (Figure 2A). The expression of dectin-1 in RAW264.7 cells was also confirmed naturally. The flow cytometry analysis after the anti-dectin-1 antibody staining showed that the RAW264.7 and d-HEK293 cells have dectin-1 on the cells (Figure 2B-D).

To examine whether the dectin-1 tarnsfectants could bind to SPG, HEK293 and d-HEK293 cells were incubated with various concentrations of b-SPG, and the binding was analyzed by using streptavidin-Alexa488 for fluorescence staining. The dectin-1 transfectants showed increased fluorescence intensity with a higher concentration of b-SPG (Figure3). However, control HEK293 cells did not show the increased binding of b-SPG at even 300 µg/ml. These

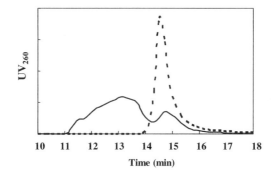

76

Figure 4. Gel permeation chromatograms of AS-ODNs (dotted line) and the reaction mixture of AS-ODNs and SPG (solid line). Chromatograms detected by UV_{260} are shown.

results indicate that b-SPG indeed bind to dectin-1 in a dose-dependent manner. Next, we investigated the uptake of SPG by using RAW264.7 cells which express dectin-1 naturally (Figure 3). RAW264.7 cells were incubated with Alexa-488 labeled SPG at 37 °C or on ice for 30 min, and after washing, the cells were analyzed by flow cytometry. As the cells were incubated on ice, their fluorescence intensity increased slightly. However, the fluorescence intensity of the cells increased drastically in the case of incubating at 37 °C. These results indicate that SPG was incorporated into the cells at 37 °C, since endocytosis is abrogated below 10 °C[16].

We prepared the complex consisting SPG and AS-ODNs having poly $(dA)_{60}$, which was introduced to conjugate with SPG. Figure 4. shows the representative GPC profile of AS-ODNs before and after the complexation with SPG. After the reaction, the AS-ODNs peak detected byUV was shifted to short elution time side, suggesting an increasing in molecular weight due to complexation of AS-ODN with SPG (solid line and dashed line). The chromatogram clearly indicated presence of unreacted AS-ODNs which eluted at 14-16 min. By measuring the area of the unreacted AS-ODN, the rate of the complexation was confirmed about 75%.

To investigate whether the AS-ODN/SPG complexes were recognized by dectin-1 on the cells, RAW264.7 cells were incubated with the b-SPG and the same volume of unlabeled SPG or complex with continuous gentle shaking, and after being washed, the incubated cells were analyzed by flow cytometry. Both the unlabeled SPG and AS-ODN/SPG complex inhibited the binding of the b-SPG to the cells (Figure 5). These data indicate that AS-ODN/SPG complex could be recognized by dectin-1. However, the unlabeled SPG inhibited the uptake of b-SPG more efficiently than AS-ODN/SPG complex.

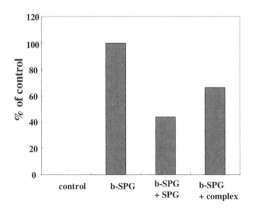

Figure 5. The recognition of SPG/ODNs complex by the cells having dectin-1. The RAW264.7 cells were treated with b-SPG and non-labeled SPG or complexes at 37 °C with gentle shaking for 30 min. The amount of b-SPG bound to the cells was determined by staining with streptavidin-Alexa 488 conjugate and using a flow cytometer.

CONCLUSIONS

In this study, we prepared the biotin-labeled SPG and evaluated the specificity to the dectin-1. By using the dectin-1 transfectants, it was confirmed that SPG was recognized by dectin-1 selectively, and indicated the uptake of SPG into the cells. We prepared the complex consisting of SPG and AS-ODNs and the resulting complex was characterized by GPC. The rate of ODN in the complex was confirmed about 75% by GPC analysis. The AS-SPG complex could inhibit the binding of b-SPG to the cells having dectin-1, suggesting that the complex can be recognized by decitn-1. These data indicate that it is possible to deliver polynucleotides into antigen presenting cells selectively by using SPG. These finding indicate that β-1,3-glucans are very attractive and useful materials in biotechnology. So we believe this novel complex can broaden the horizon of a gene delivery technology.

REFERENCES

[1]   N.M. Dean and C.F. Bennett, Oncogene 22 (2003) 9087-96.
[2]   S.T. Crooke, Annu Rev Med 55 (2004) 61-95.
[3]   J.H. Chan, S. Lim and W.S. Wong, Clin Exp Pharmacol Physiol 33 (2006) 533-40.
[4]   M. Meyer, A. Philipp, R. Oskuee, C. Schmidt and E. Wagner, J Am Chem Soc 130 (2008) 3272-3.
[5]   M. Grzelinski, B. Urban-Klein, T. Martens, K. Lamszus, U. Bakowsky, S. Hobel, F. Czubayko and A. Aigner, Hum Gene Ther 17 (2006) 751-66.
[6]   J. DeRouchey, C. Schmidt, G.F. Walker, C. Koch, C. Plank, E. Wagner and J.O. Radler, Biomacromolecules 9 (2008) 724-32.
[7]   H. Kang, R. DeLong, M.H. Fisher and R.L. Juliano, Pharm Res 22 (2005) 2099-106.
[8]   H. Lv, S. Zhang, B. Wang, S. Cui and J. Yan, J Control Release 114 (2006) 100-9.
[9]   S. Akhtar and I.F. Benter, J Clin Invest 117 (2007) 3623-32.
[10] K. Sakurai and S. Shinkai, J. Am. Chem. Soc. 122 (2000) 4520-4521.
[11] K. Sakurai, M. Mizu and S. Shinkai, Biomacromolecules 2 (2001) 641-50.
[12] T. Norisuye, T. Yanaki and H. Fujita, J. Polym. Sci., Polym. Phys. Ed. 18 (1980) 547-558.
[13] T. Sato, T. Norisuye and H. Fujita, Macromolecules 16 (1983) 185-189.
[14] G.D. Brown and S. Gordon, Nature 413 (2001) 36-7.
[15] G.D. Brown, P.R. Taylor, D.M. Reid, J.A. Willment, D.L. Williams, L. Martinez-Pomares, S.Y. Wong and S. Gordon, J Exp Med 196 (2002) 407-12.
[16] W.A. Dunn, A.L. Hubbard and N.N. Aronson, Jr., J Biol Chem 255 (1980) 5971-8.

Mater. Res. Soc. Symp. Proc. Vol. 1140 © 2009 Materials Research Society   1140-HH06-28-DD03-28

Cytotoxicity and Biological Effects of Functional Nanomaterials Delivered to Various Cell Lines

Meena Mahmood*, Enkeleda Dervishi, Yang Xu, Zhongrui Li, Mustafa Abd Al-Muhsen, Nawab Ali, Morgan Whitlow, and Alexandru S. Biris*

Applied Science Department, Nanotechnology Center, University of Arkansas at Little Rock, 2801 South University Avenue, Little Rock, AR 72204, Fax: 501-683-7611.
*Corresponding authors Emails: mwmahmood@ualr.edu, asbiris@ualr.edu

Nanostructured materials have been found to be uptaken by various cell lines and highly affect their biological behavior. In this work, gold and silver metal nanoparticles as well as single wall carbon nanotubes were incubated separately and with apoptotic agents (Dexamethasone and Etoposide), in cell cultures of mouse long murine osteocytic bone cells (MLO-Y4 cells) and human cervical cancer cell line (Hela cells). The incubation of the nanomaterials with the cell cultures was carried out at two concentrations (0.5 X 10^{-9} mol/L and 10^{-12} mol/L) for 24 hours and the apoptotic agents (10^{-5} mol/L and 75 X 10^{-6} mol/L) were introduced for six hours. The cytotoxicity data revealed that the Au-NPs had a significant lower cytotoxic effect than the Ag-NPs and CNTs, values reflected by the percentage of the dead cells vs. living cells and that the combination of the nanomaterials and the apoptotic agents had a combinatorial effect, resulting in a significantly larger number of cells that died. The results highlighted by this study could represent a major development for the delivery of specific drug molecules into cancer cells and tumors by nanomaterials.

INTRODUCTION

Some of the most promising applications include structural engineering, electronics, optics, consumer products, alternative energy, soil and water remediation, or for medicinal uses as therapeutic, diagnostic or drug delivery devices[1]. The promising field of nanomedicine offers the potential of monitoring, repair, construction and control of human biological systems at the molecular level[1, 2] and has resulted in the engagement by drug companies in nanotechnology research. New thinking is required not only in understanding toxicology associated with nanomaterials, but in the understanding of all toxicants to which the human is exposed[3].

Given the unique dimensional and morphological properties of the nanomaterials, a large number of major applications that have been developed and which hold significant promise in the successful targeting of cancer[4], tumor ablation[5], drug and gene delivery[6], and especially tissue engineering[7]. In the last few years a large number of research publications have indicated that the nanomaterials have the ability to interact very strongly with various biological systems. Various types of cell lines have been shown to grow on nano-based substrates such as carbon nanotubes or other nanomaterials[8]. Moreover the strong interaction of such biological systems with nanomaterials, it is also well known to induce an uptake of nanomaterials inside various sub-cellular components and tissues[9]. Therefore, a more thorough understanding of the potential cytotoxic effects of such nanomaterials is required. This work presents the morphological cellular changes that are induced by the uptake by bone and cancer cells of various nanomaterials. Moreover, it was found that the nanoparticles have the ability to enhance the effects of several commonly used apoptotic agents.

MATERIALS AND METHODS:

NANOPARTICLES SYNTHESIS
Silver NPs of 10 - 20 nm diameter were prepared by citrate or borohydride mediated reduction of chloroauric acid or silver nitrate, respectively based on the following protocol: In DI water introduce sodium borohydrate followed by sodium citrate followed by AgNO2 (drop wise) under slow stirring. Immediately add PVP was to the solution and let stir for 30 minutes. Product should be a golden yellowish in color. Gold nanoparticles 20 nm was purchased from (Sigma Aldrich Company). SW-CNTs were sonicated with the growing medium freshly just before culturing and the concentration of the nanoparticles was determined by UV –Vis / AFM techniques.

CELL CULTURING:
MLO-Y4 Osteocytic cells obtained from murine long bone, and Human Cervical cancer cells Hela were grown 75 cm^2 flasks with minimum essential medium supplemented by 10% fetal bovin serum and 1 % penicillin, streptomycin and gentamycin antibiotics (4 mM L-glutamine and 100 U/ml of each penicillin and streptomycin), For the primary culture; the cells were cultured in humidified incubator at $37^{\circ}C$ and 5% CO_2 for few days.

NANOMATERIALS ADMINSTRATION:
Silver NPs, Gold NPs and CNTs were delivered to the cells by fluid uptake in a concentration of (0.5 X 10^{-9} mol/L and 10^{-12} mol/L). Cell lines were trypsinized and transferred to the 48 well plates in a desired density of 10^4/well and incubated overnight under the same condition with the nanomaterials.

APOPTOSIS INDUCTION:
MLO-Y4 Osteocytic cells were used for apoptosis induction assay. (10^{-5} mol/L and 75 X 10^{-6} mol/L) of Dexamethasone, Etoposide, and cell culture vehicle only (ethanol and DMSO) were separately administered to the cell cultures and with the combination of nanomaterial solutions (Au-NPs, Ag-NPs, SWCNT). Incubation for 6 hrs was performed. Experiment also included a 3 control samples with exposure to only cell culture vehicle, Dexamethasone or Ethoposide, respectively.

CELLS VIABILITY ANALYSIS AND TRYPAN BLUE ASSAY:
The viable cells percentages were measured by Trypan Blue dye to stain the cells that have an intact membrane. First, both cell lines were cultured in a 48-well plate in a desired density under the same conditions. Then, the cells were dissociated from the bottom of the plate by trypsin and transferred to 1.5 eppendorf tubes and spin dawn. Finally, 25 μL of 1 X Trypan Blue dye was added to each sample and incubated for less than 5 min. The viable cells number was counted through a hemocytometer, and the viability values were derived by the following equation and comparing the samples with the negative control.
Nonviable cells (%) = (Number of nonviable cells X100)/No. of viable cells + No. of nonviable cells

ACTIVE CASPASE-3 ASSAY:

Caspase-3-like assay was done by the aid of the Biovision caspase-3 assay kit. Briefly, the apoptosis was induced with the desired methods, the cells were collected by the cell scraper and transferred to the 1.5 eppendorf tubes and incubated with 1 μl of the Red-DEVD-FMK and incubated for 1 hour at $37°C$ with 5% CO_2, and spin dawn for 5 minutes at 3000 rpm and the supernatant removed carefully, the cells resuspended with 50 μl of the washing buffer a and spin dawn again; finally the cells resuspended with 100 μl of the washing buffer and few drops of the cell suspension were transferred to the microscopic slides and the brightness of the red stain measured. The brightest red cells have the active caspase - 3 while the less red stained cells have the less activated caspase-3.

STATISTICAL ANALYSIS:

All data were expressed as mean ± std. dev. Differences among 3 or more groups were evaluated by means of 1- way-ANOVA test and independent- sample- t -test was performed for 2 group comparisons. P values of 0.05 or less were considered to indicate significance.

RESULTS AND DISCUSSION:

EFFECT INDUCED BY NANOMATERIALS ADMINSTRATION:

The nanoparticles in various concentrations were delivered to the MLO-Y4 cell line. Culture medium suspensions containing 0 mol/L (vehicle-only), 0.5×10^{-9} mol/L as well as of 10^{-12} mol/L nanomaterial concentration were prepared for each nanomaterial. Apoptotic rate was calculated for each sample.

Administration of all three types of nanomaterial induced increased apoptotic cells rate level compared to vehicle samples, as shown in Table 1. Results also indicate significant differences for 2 group comparisons between 0.5×10^{-9} mol/L suspensions and 10^{-12} mol/L suspensions containing the same type of nanomaterial (p<0.05 for each material).

Table 1. Apoptotic cell rate at different concentrations for each nanomaterial type.

	Apoptosis Rate (%)			
	Medium only (0 mol/L concentration)	High nanomaterials concentration $(0.5 \times 10^{-9}$ mol/L)	low nanomaterials concentration $(10^{-12}$ mol/L)	P *
Au-NP	2.63±0.56	3.02±0.87	4.10±0.38	<0.05
Ag-NPs	2.63±0.56	4.72±0.64	6.90±0.89	<0.05
SW-CNTs	2.63±0.56	7.81 ±1.10	11.58 ±2.06	<0.05

* p values were calculated by mean of 1- way ANOVA test across the three concentration samples .

EFFECT INDUCED BY NANOMATERIAL PHYSICAL AND CHEMICAL PROPERTIES:

Table (1) also illustrates differences among nanomaterial effects at the same concentration level. Apoptotic effect induced by low concentration suspensions (10^{-12} mol/L) was highest in SW-CNTs (7.81 ±1.10), followed, in descending order by Ag-NP (4.72±0.64) and Au-NP (3.02±0.87). Differences between any two groups were found to be significant (p<0.05). High concentration suspensions (0.5×10^{-9} mol/L) induced the same apoptotic rate pattern for the three groups (11.58 ±2.06, 6.90±0.89 and 4.10±0.38 for CNTs, Ag-NPs and Au-NPs, respectively). Similar to low concentration suspensions results, p values were significant for comparison between any two groups (p<0.05).

INDIVIDUAL AND COMBINED EFFECT OF NANOMATERIALS AND ANTIPROLIFERATIVE AGENTS:

Two already established apoptotic agents were tested in relationship with nanomaterials: dex. (D) and etoposide (E). Each nanomaterial was separately delivered to the MLO-Y4 cell line in the presence as well as in the absence of D and E. Controls consisted of vehicle-only (Ethanol+ DMSO) administered cell culture samples.

Results are presented in Table 2. Significantly higher apoptotic effects were recorded for combined nanomaterial – apoptotic agent samples compared to simple nanomaterial or D/E administrated samples.

Table 2. Results of cell line exposure to combined apoptotic agents.

Solutions	Apoptotic rate (%)			
		Nanomaterials		
	Control	Au-NPs-added sollution	Ag-NPs-added sollution	SW-CNTs-added sollution
Vehicle (Ethanol+DMSO)	2.25 ± 0.44	3.42 ± 0.41	6.58 ±0.58	7.46 ± 0.73
Dexamethasone (D+vehicle)	4.47 ± 0.54	10.04 ± 1.10	12.85 ±0.34	29.43 ± 0.78
Etoposide (E+vehicle)	7.64 ± 0.42	13.82 ± 0.70	19.60 ±0.97	40.37 ± 0.81

Table 3. Individual and combined apoptotic cell rates of nanomaterials and Dexamethasone/Etoposide.

Simple Effect		Additive Effect		Real Effect		Increase of Additive Effect	
		D-added samples	E-added samples	D-added samples	E-added samples	D-added samples	E-added samples
Au-NPs	1.17±0.44	3.39±0.44	4.34±0.42	7.79±1.10	11.57±0.70	129.79%	166.58%
Ag-NPs	4.33±0.58	6.55±0.54	7.50±0.58	10.60±0.34	17.35±0.97	61.83%	131.33%
SW-CNTs	5.21±0.73	7.43±0.73	8.38±0.42	27.18±0.78	38.12±0.81	265.81%	354.89%
Dexamethasone	2.22±0.54						
Etoposide	3.17±0.42						

As shown in Table 3 the apoptotic effect of both D and nanomaterial added groups were significantly larger than the simple additive effect exerted by dex. and nanomaterial alone (p<0.05 for each type of nanomaterial). Maximal increase of summative effect was obtained for SW-CNTs (265.81%), followed by Au-NPs (61.83%) and Ag-NPs (129.79%). Groups that were exposed to both E and each type of nanomaterial demonstrated similar increased response and cytotoxicity hierarchy. Increase of additive effect was found to be significantly higher in Etoposide administered samples (p<0.05 for each nanomaterial).

APOPTOTIC AGENTS AND CASPASE-3 ACTIVATION MECHANISM:

We explored the use of the active form of caspase-3 for the detection of early apoptotic events. The Hela cells were treated with different nanomaterials with and without the combination with of the apoptotic agents and labeled with Red-DEVD-FMK Caspases 3 staining kit as described earlier. The significance of the caspase-3 activation shown by the quantification of the red stain in each sample by the fluorescent microscope and the brighter red shows more cleaved caspase-3. The combination of the SW-CNTs with the Etoposide shows the most significant activation and approximately most of the cells appear with bright red stain while the other treatments show the lower brightness of the red stain with lower caspase-3 activation level and the non-apoptotic cells appear with dark background. Like in image (I, J and K) we can see so many cells having the bright red stain in comparison with the other images that means the canbon nanotubes cleaved more active caspase-3, also we can compare within each variable so the cells were cultured with combination of the nanomaterials and the apoptotic agents have the most quantity of the red stain in comparison with the vehicle samples. In this present work we used mammalian cervical cancer cells (Hela Cells) to detect the active caspase-3 as an apoptotic marker.

Our research also brings evidences of nanomaterial activation potential for caspase-3 apoptotic pathway. The protease is responsible for the initiation of the death cascade and is therefore an important marker of the cells entry point into the apoptotic signaling pathway [10]. Caspase-3, one of 13 aspartate-specific cystein proteases that play an important role in the execution of the apoptosis program; which is primarily responsible for the cleavage of PARP during cell death which is lead to the degradation and the fragmentation of the chromosomal DNA inside the nucleus and apoptosis of the cell. [11, 12]

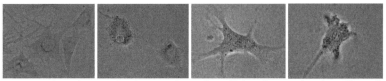

Figure 1: The microscopic pictures show the cellular changes by the effect of the nanomaterials and the apoptotic agents, (A) cells were cultured with silver NPS overnight and treated with higher conc. of Etop. (B) Nuclear Chromatin condensation and cellular membrane blabbing. (C) Cellular shrinkage. (D) Cell lysis and cellular disintegration by the comparison with alive cell.

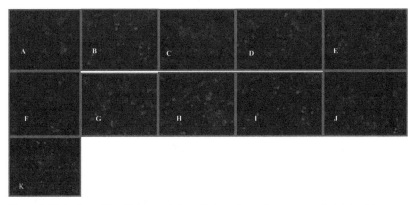

Figure 2: Detection of the Hela apoptotic cells by using active caspase-3 staining kit. (A) Hela cells were cultured with the appropriate growth medium as a control, (B)the cells were cultured with vehicle (ethanol and DMSO), (C) the cells were cultured with gold NPS, (D)gold NPS with dex., (E)gold NPS with etopo., (F)silver NPS, (G)silver NPS with dex., (H)silver NPS with etopo., (I)SW-CNTs, (J)SW-CNTs with dex., and (K)SW-CNTs with etopo.

CONCLUSIONS

Nanomaterials alone and (more importantly) in combination with classical antiproliferative agents such as: Dexamethasone and Etoposide induce apoptosis by activation of caspase-3 pathway and have a high potential for use as chemotherapy. Further research is needed for generating new methods of manipulating their physical and chemical properties for optimising efficiency and decreasing possible side effects.

References

[1]. Oberdorster G, Oberdorster E. 2005b. Environ Health Perspect 133(7):823-839.

[2]. Borm PJ, Robbins D, Haubold S, Kuhlbusch T, Fissan H, Donaldson K, Schins R, Stone V, Kreyling W, Lademann J. 2006. Part Fibre Toxicol 3:11.

[3]. NRC, Toxicity Testing in the 21st Century: a Vision and a Strategy, 2007, Washington DC: National Academic Ptress.

[4]. Joe EK, Wei X, Anderson RR, and Lin CP. Biophysical J. 2003;84:4023-4032.

[5]. Zharov V, Galitovsky V, and Viegas M. Appl Phys Letter 2003; 83(24):4897-4899.

[6]. Jayanth Panyam· and Vinod Labhasetwar. Volume 55, Issue 3, 24 February 2003, Pages 329-347

[7]. Benjamin S. Harrison, and Anthony Atala, Volume 28, Issue 2, January 2007, Pages 344-353

[8]. Laura P. Zanello, Bin Zhao, Hui Hu, and Robert C. Haddon, 2006 Vol. 6, No. 3, 562-567.

[9]. Nadine Wong Shi Kam, Michael O'Connell, and Hongjie Dai, PNAS, 11600–11605 2005, vol. 102 no. 33.

[10]. Nicholson DW, Ali A, Ding CK, Gallant M, Gareau Y, Griffin PR, Labelle M, Lazebnik YA, 1995 Jul 6; 376(6535):37-43.

[11]. S Nagata, H Nagase, K Kawane, N Mukae and H Fukuyama.Cell Death and Differentiation (2003) 10, 108–116.

[12]. A. Hamid Boulares, Alexander G. Yakovlev, Vessela Ivanova, Sudha Iyer, and Mark Smulson. J Biol Chem, Vol. 274, Issue 33, 22932-22940, August 13, 1999.

Mater. Res. Soc. Symp. Proc. Vol. 1140 © 2009 Materials Research Society   1140-HH03-05

Nanoparticle-Tagged Perfluorocarbon Droplets for Medical Imaging

Naomi Matsuura[1], Ivan Gorelikov[1], Ross Williams[1], Kelvin Wan[1], Siqi Zhu[1], James Booth[2], Peter Burns[1,3], Kullervo Hynynen[1,3] and John A. Rowlands[1,3]

[1]Imaging Research, Sunnybrook Health Sciences Centre, Toronto, Ontario, Canada M4N 3M5

[2]Molecular and Cellular Biology, Sunnybrook Health Sciences Centre and Department of Immunology, University of Toronto, Toronto, Ontario, Canada M4N 3M5

[3]Department of Medical Biophysics, University of Toronto, Toronto, Ontario, Canada M4N 3M5

ABSTRACT

Ultrasound-activatable, nanoparticle-tagged perfluorocarbon (PFC) droplets were developed for medical imaging. Silica-coated QD nanoparticles were dispersed in PFCs and subsequently emulsified to form sub-micron, nanoparticle-tagged PFC droplets in water. Fluorescence and transmission electron microscopy were used to confirm that the nanoparticle-tagged PFC droplets successfully labeled live macrophage cells *in vitro*. The nanoparticle-tagged PFC droplets were activated to disperse the nanoparticles using an ultrasound beam, which converted the droplets to gas bubbles that were rapidly driven to collapse. At ultrasound exposure pressures greater than 1.7 MPa, the activation of the PFC droplets in the cells resulted in a >90% decrease in the total cell count of PFC-loaded cells. This work reveals the potential of using ultrasound-activatable PFC constructs containing medical nanoparticles to image and treat diseased cells *in vivo* with ultrasound.

INTRODUCTION

There is interest in designing multifunctional architectures for clinical medical imaging that can both amplify weak native contrast and deliver localized therapy at disease sites [1-3]. Silica-coated nanoparticles have been proposed as both medical imaging and therapy agents [4-6], but existing *in vivo* biological barriers often limit the ability of nanoparticles to reach their target sites [7]. The development of constructs that are designed to carry nanoparticles within their cores and can be remotely activated by an energy source external to the patient may result in precisely controlled nanoparticle dispersion to a highly localized and specific target area. Perfluorocarbon (PFC) droplets have been shown to be effective medical imaging contrast agents and radiosensitizers [8-13], and can be converted from the liquid phase to gaseous bubbles through an ultrasound-mediated phase transition [10, 14]. In the present study, the integration of silica-coated quantum-dot (QD) nanoparticles into PFC droplets that can be activated with ultrasound to locally release their nanoparticle payloads was investigated for multifunctional imaging/therapy applications under *in vitro* conditions.

EXPERIMENTAL DETAILS

Silica-coated QD nanoparticles were synthesized according to a previous method [15] and coated with silica [16, 17]. For dispersion of the silica-coated nanoparticles into PFCs (3M), they were reacted with perfluorodecyltriethoxysilane (Sigma-Aldrich) [18]. Nanoparticle-labeled PFC droplets-in-water emulsions were prepared using deionized water, nanoparticle-PFC solution, and fluorosurfactant (Sigma-Aldrich) using sonication (Misonix Sonicator 3000). After emulsification, the nanoparticle-tagged PFC droplets were resuspended in PBS. The fluorescence spectra of the emulsions were obtained using a Horiba Jobin Yvon FluoroMax-4 spectrofluorometer. The hydrodynamic diameters of the PFC droplets were obtained using a Malvern Nanosizer S instrument.

For *in vitro* studies, nanoparticle-labeled PFC droplets were incubated with murine alveolar macrophage cells (RAW264.7, ATCC (Rockville, MD)), maintained at 37°C in a humidified atmosphere with 5% CO_2 in DMEM medium (Wisent) supplemented with 10% fetal bovine serum (Wisent). After incubation, the cells were washed with sterile PBS to remove excess PFC droplets. Qualitative uptake of nanoparticles was analyzed by inverted epifluoresence microscopy using a Zeiss Axiovert 200M microscope with Axiovision imaging software and standard FITC, cy3, and DAPI filter sets. Size distributions of cells were measured with a Coulter counter (Multisizer III, Beckman Coulter Inc, Fullerton, CA). A 750 µL volume of the PBS-diluted cell suspension was re-diluted in 10 mL Isoton-II electrolytic solution. A 50 µm-diameter aperture was used to count the number of cells with diameters between 1-30 µm in a 100 µL sample volume.

Ultrasound activation of the nanoparticle-labeled PFC droplets was performed using the apparatus depicted in **figure 1**. The ultrasound experiments were repeated three times. For each experiment, the labeled and unlabeled cells were split into 2 mL aliquots, with each exposed for 90 s to 80 µs ultrasound bursts at 1 MHz, repeated every millisecond, at peak-negative pressures ranging from 0-2 MPa. The cells were placed in a 1-cm-diameter exposure chamber with Mylar windows, situated at the focus of the ultrasound beam and immersed in a water bath. An arbitrary waveform generator (AWG 520, Tektronix, Richardson, TX) was used to generate the ultrasound pulses, which were amplified (RPR4000, Ritec Inc, Warwick, RI) and transmitted through a 1 MHz circular single-element transducer with 37 mm aperture and 75 mm focal length (IL0112HP, Valpey Fisher, Hopkinton, MA). Received echoes originating from inertial cavitation of PFC gas bubbles [19] were detected with a 500 kHz circular single-element transducer with 37 mm aperture and 75 mm focal length (IL0512HP, Valpey Fisher, Hopkinton, MA), and were recorded using a 12-bit digitizer (DC440, Acqiris, Geneva, Switzerland) at 100 MS/s, triggered by the AWG. Measurements with a calibrated membrane hydrophone (804, Sonora Medical Systems Inc, Longmont, CO) were performed to determine the peak negative pressures at the focus of the 1 MHz transducer.

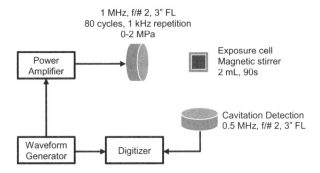

1 MHz, f/# 2, 3" FL
80 cycles, 1 kHz repetition
0-2 MPa

Power Amplifier

Exposure cell
Magnetic stirrer
2 mL, 90s

Cavitation Detection
0.5 MHz, f/# 2, 3" FL

Waveform Generator

Digitizer

Figure 1. Schematic of the ultrasound apparatus used for the activation of the QD-labeled PFC droplets and detection of the activation event.

RESULTS AND DISCUSSION

Silica-coated nanoparticles were incorporated into PFC droplets, and were incubated with macrophage cells for maximal uptake. To determine if the PFC-loaded cells could be activated upon exposure to ultrasound, acoustic emissions indicative of cavitation [19] were measured during ultrasound exposures at various pressures. The activation of the nanoparticle-tagged PFC droplets within cells was verified through measurements of the size distributions of the cells before and after exposure to ultrasound.

The successful loading of nanoparticle-tagged PFC droplets into cells was determined using silica-coated CdSe/ZnS core/shell QD nanoparticles combined with fluorescent microscopy. In this way, the QD fluorescence could verify nanoparticle-tagging of the PFC droplet after synthesis, nanoparticle-tagged PFC loading into cells, and the fate of the PFC droplets and nanoparticles after activation. The QD nanoparticles within the PFC droplets exhibited emission peaks at 610 nm (FWHM <30 nm), and no shift in their photoluminescence spectra was detected as compared to the original silica-coated QDs in water. Before incubation with cells, the QD-labeled PFC droplets were homogeneously distributed in water, with droplet diameters measured to be ~200 nm. No difference in size between QD-labeled PFC droplets and their non QD-labeled PFC droplet counterparts was observed. The incubation of QD-tagged PFC droplets with macrophage cells *in vitro* for two hours resulted in significant loading of the droplets into cells, as observed using fluorescence microscopy. Transmission electron microscopy of fixed QD-tagged PFC droplets in cells also showed the QD labels in vacuoles within the cell, which was confirmed by energy dispersive X-ray spectroscopy. Extended incubation (>24 hours) of QD-tagged PFC droplets with cells did not result in a decrease in cell count as compared to unlabeled cells.

To evaluate if QD-labeled PFC droplets in cells could be activated by ultrasound [20, 21], experiments were conducted on macrophage cells loaded with QD-tagged PFC droplets, with negative control experiments performed on cells which did not contain PFC droplets. The suspended cells were diluted in PBS to a concentration of approximately 2×10^6 per mL and were kept at 37°C before and after each ultrasound exposure. The effect of ultrasound on macrophages which contained QD-tagged PFC droplets was demonstrated to be pressure dependent (**figure 2**). The successful QD-tagged PFC droplet activation was detected with ultrasound in real-time using the enhanced acoustic emission resulting from cavitation. Inertial cavitation is characterized by broad-band acoustic emissions [19]. A small amplitude signal was scattered by the chamber at a pressure of 1.4 MPa (**figure 2a**). This signal was static during each exposure and was detected because of overlap between the bandwidths of the 500 kHz and 1 MHz transducers. As the pressure was increased from 1.6 to 1.8 MPa, a threshold was observed above which a large, rapidly fluctuating signal was detected. At 1.7 MPa, corresponding to the onset of cellular disruption, a signal was detected intermittently. Above the threshold, cavitation events occurred continuously and a large amplitude signal was detected (**figure 2b**). A characteristic snapping sound associated with microbubble cavitation [19] was also audible at incident pressures which exceeded the threshold. For the control experiments, only the small signals scattered by the chamber were detected; no cavitation signals were evident at pressures up to 2.0 MPa.

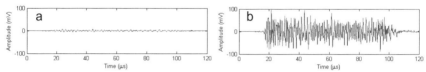

Figure 2. Ultrasound signals detected from the PFC-loaded cells during exposure at ultrasound pressures: (a) below the activation threshold (1.4 MPa); and (b) above the activation threshold (2.0 MPa).

The sensitivity to which the cells were destroyed by the activation of the nanoparticle-tagged PFC droplets can be observed in the abrupt change in the total cell count as a function of pressure (**figure 3**). The size distribution of the ultrasound-exposed cell samples as measured with a Coulter counter was found to be consistent with activation as detected by inertial cavitation measurements. The average cell diameter was 11.2 ± 1.8 μm. The distribution was virtually unchanged for exposures up to 1.5 MPa, compared with the unexposed samples. At pressures ranging from 1.5-1.8 MPa, the total cell count between 8-16 μm decreased substantially (**figure 3**), indicating fragmentation of the cells. At 2.0 MPa, no cells remained, and a negligible amount of fragmentary particles greater than 5 μm in diameter were detected. For the control experiments with unlabeled cells, there were no significant changes in the size distributions for pressures up to 2.0 MPa. Although the experiments were done as quickly as possible and with gentle handling to minimize cell death, a decrease of up to 10% in the total cell count was observed for the lowest-pressures exposures relative to the unexposed controls, likely due to the natural fragility of the cells.

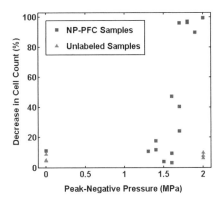

Figure 3. Decrease in total cell count of QD-labeled PFC-loaded cells and unlabeled-unloaded cells exposed to ultrasound at different pressures.

CONCLUSIONS

This study shows for the first time that nanoparticle-tagged PFC droplets may be synthesized, loaded into cells, and activated using an ultrasound beam, causing the release of the nanoparticle payload and cell death, and potentially enabling the imaging and treatment of diseased cells *in vivo*. The modular structure and the simple substitution of the QDs with other silica-coated medical imaging or therapy nanoparticles [4-6] within the already multimodal PFC encapsulant can result in user-defined, multifunctional imaging and therapy agents, mitigating the individual limitations of current single-purpose agents. This type of composite, hierarchical construct may allow flexible, multifunctional agents for specific imaging and therapeutic applications to be rapidly developed through the modular assembly of pre-validated components.

ACKNOWLEDGMENTS

We are grateful to Dr. Philippe Poussier (Sunnybrook Research Institute) for discussions regarding cell models for PFC droplet testing. This study was conducted with the support of the CIHR Excellence in Radiation Research for the 21st century (EIRR21st) Research Training Program, the Ontario Institute for Cancer Research Network through funding provided by the Province of Ontario, the US FY07 Department of Defense Breast Cancer Research Program Concept Award and a grant from the Terry Fox Foundation.

REFERENCES

1. N. Matsuura and J.A. Rowlands, *Med. Phys.* **35**, 4474 (2008).
2. M. Ferrari, *Nat. Rev. Cancer* **5**, 161 (2005).
3. K. Y. Kim, *Nanomedicine* **3**, 103 (2007).
4. C. Zhang, B. Wangler, B. Morgenstern, H. Zentgraf, M. Eisenhut, H. Untenecker, R. Kruger, R. Huss, C. Seliger, W. Semmler and F. Kiessling, *Langmuir* **23**, 1427 (2007).
5. I. I. Slowing, J.L. Vivero-Escoto, C.-W. Wu and V. S.-Y. Lin, *Adv. Drug Del. Rev.* **60**, 1278 (2008).
6. J. Kim. H. S. Kim, N. Lee, T. Kim, H. Kim, T. Yu, I. C. Song, W. K. Moon and T. Hyeon, *Angew. Chem. Int. Ed.* **47**, 8438 (2008).
7. S. M. Moghimi, A. C. Hunter, and J. C. Murray, *Pharmcological Reviews*, **53**, 283 (2001).
8. Y.B. Yu, *J. Drug Target.* **14**, 663 (2006).
9. S. Rockwell, M. Kelley, C. G. Irvin, C. S. Hughes, E. Porter, H. Yabuki, and J. Fischer, *Radiother Oncol*, **22**, 92 (1991).
10. R.F. Mattrey, F. W. S. Scheible, B. B. Gosink, G. R. Leopold, D. M. Long, and C. B. Higgings, *Radiology* **145**, 759 (1982).
11. M.-S. Liu and D.M. Long, *Radiology* **122**, 71 (1977).
12. A. M. Morawski, G. A. Lanza, and S. A. Wickline, *Curr. Opin. Biotechnol.*, **16**, 89 (2005).
13. G. M. Lanza, and S. A. Wickline, *Progress in Cardiovascular Diseases* **44**, 13 (2001).
14. K. Kawabata, N. Sugita, H. Yoshikawa, T. Azuma, and S. Umemura, *Jpn. J. Appl. Phys* **44**, 4548 (2005).
15. J. J. Li, Y. A. Wang, W. Guo, J. C. Keay, T. D. Mishima, M. B. Johnson, and X. Peng, *J. Am. Chem. Soc.* **125**, 12567 (2003).
16. D. K. Yi, S. T. Selvan, S. S. Lee, G. C. Papaefthymiou, D. Kundaliya, and J. Y. Ying, *J. Am. Chem. Soc.* **127**, 4990 (2005).
17. I. Gorelikov and N. Matsuura, *Nano Letters* **8**, 369 (2008).
18. N. Matsuura, presented at the 2008 World Molecular Imaging Conference, Nice, France, 2008 (unpublished).
19. T. G. Leighton, *The Acoustic Bubble*. San Diego: Academic Press (1994).
20. T. Giesecke, and K. Hynynen, *Ultrasound Med. Biol.* **29**, 1359 (2003).
21. A. H. Lo, O.D. Kripfgans, P.L. Carson, E.D. Rothman, and J.B. Fowlkes, *IEEE Trans. Ultrason., Ferroelect., Freq. Contr.* **54**, 933 (2007).

Mater. Res. Soc. Symp. Proc. Vol. 1140 © 2009 Materials Research Society   1140-HH05-11

Novel Superparamagnetic Iron Oxide Nanoparticles for a Multifunctional Nanomedicine Platform

O. Taratula[1,3], R. Savla[1,3], I. Pandya[1], A. Wang[2], T. Minko[3,4], H. He[1,4]
[1] Department of Chemistry, Rutgers University, 73 Warren Street, Newark, NJ 07102, U.S.A.
[2] Ocean NanoTech, Fayetteville, AR 72701, U.S.A.
[3] Department of Pharmaceutics, Ernest Mario School of Pharmacy, Rutgers, The State University of New Jersey, Piscataway, NJ 08854, U.S.A.
[4] Cancer Institute of New Jersey, 195 Little Albany Street, New Brunswick, NJ 08903, U.S.A.

ABSTRACT

The rational for this study is to develop a multifunctional nanomedicine platform for siRNA delivery to cancerous tumors as well as to use magnetic resonance imaging (MRI) as a noninvasive strategy to monitor the drug therapy outcome. Therefore, we propose a multifunctional gene delivery vector by combining the magnetic properties of superparamagnetic iron oxide (SPIO) nanoparticles and their ability to package siRNAs into discrete complexes. The outer surface of SPIO was modified with two different polymer layers in tandem. The new design of nanoparticle surface modification allows sufficient binding strength to package large number of siRNAs and protect them in the delivery route. It also facilitates the siRNAs escape from endosomes to cytoplasm where they enter RNAi pathway. The prepared nanoparticles demonstrate high efficiency to provoke siRNAs complexation and largely facilitate their transfection into cancer cells. Moreover, to develop a specifically targeted, multifunctional siRNA delivery system with high transfection efficiency, the ability of SPIO nanoparticles and poly(propyleneimine) generation 5 dendrimer (PPI G5) to cooperatively provoke siRNA complexation was employed in the current study. In order to improve the efficiency of the designed vector, poly(ethylene glycol) (PEG) and receptor-binding ligands have been incorporated into SPIO/PPI G5/siRNA complexes to enhance serum stability and selective cellular internalization. The modified siRNA nanoparticles can sufficiently enhance the targeted cellular internalization and reduce the expression of the specific genes.

INTRODUCTION

RNA interference (RNAi) therapeutics represents a fundamentally new way to treat human disease by addressing targets that are otherwise untreatable with existing medicine [1]. However, overcoming the delivery obstacles both *in vivo* and *in vitro* is the greatest barrier for using short siRNAs as therapeutic drugs. Except the challenges for *in vivo* delivery, such as extracellular stability and specific targeting, the difficulty for *in vitro* delivery is the combination of the abilities of high cellular entry, quick endosomal escape, dissociation from the carrier, and coupling with cellular machines (RISC). Using engineered nanoparticles with imaging capability to delivery siRNAs will lead to the development of powerful multifunctional nanomedicine to treat cancer or other difficult diseases. Therefore, we propose a multifunctional gene delivery vector by combining the magnetic properties of SPIO nanoparticles and their ability to package siRNAs into discrete complexes [2, 3]. The different size of iron oxide nanoparticles were fabricated and their outer surface was modified with two different polymer layers in tandem such as poly (maleic anhydride-*alt*-1-octadecene) (PMAO) and positively charged poly (diallyldimethylammonium chloride) (PDDA). The new design of nanoparticle surface modification allows sufficient binding strength to package large number of siRNAs and protect them in the delivery route. It also facilitates the siRNAs escape from endosomes to cytoplasm where they enter RNAi pathway. Moreover, to specifically deliver this multifunctional nanomedicine to cancer cells as well as increase their stability in serum, siRNA complexes were also prepared with a mixture of SPIO nanoparticles and poly(propyleneimine) generation 5 dendrimer (PPI G5). The free amine groups on the surface of the formed complexes were used to conjugate a layer of polyethylene glycol (PEG). The distal end of PEG was then modified with targeting agent, such as Luteinizing hormone-releasing hormone (LHRH), which is overexpressed in many types of cancer cells and is not expressed in healthy organs [4]. The LHRH modified siRNA nanoparticles packaged with the mixture of SPIO nanoparticles and PPI G5 dendrimers are capable of targeting cellular internalization and can sufficiently reduce the expression of *BCL2* gene.

EXPERIMENTAL DETAILS

The complexes of siRNA with 5 nm and 10 nm SPIO nanoparticles, PPI G5 dendrimers, the mixtures of 5 nm SPIO with PPI G5 and 10 nm SPIO with PPI G5 were prepared at different amine/phosphate ratios (N/P ratio) such as 0.16, 1.50, 1.19, 0.73 and 1.30, respectively. In general, siRNA solution (100 μM) was mixed with either DI water or HEPES buffer (10 mM, pH 7.2) by following adding of appropriate amount of the condensing agents. The complexes were shortly vortexed and allowed to equilibrate for 30 min prior to analysis. In order to modify the formulated siRNA nanoparticles, NHS-PEG-MAL was reacted with primary amines on the surfaces of the siRNA particles in 10 mM HEPES buffer (pH 7.2). The ratio of primary amines to PEG was 35/1. The reaction was carried out for 1 hr at room temperature. Then LHRH peptide dissolved in HEPES buffer was introduced into the solution and incubated overnight at 4°C. The resulting product was dialyzed against deionized water (MWCO 10,000). The formulated siRNA nanopartices were imaged with tapping mode atomic force microscope in ambient air (Nanoscope III A, Digital Instruments).

LHRH-positive (A549) and negative (SKOV-3) cells were plated (20, 000 cells/well) in 6-well tissue culture plate and treated with the 6-Carboxy-Fluorescein (FAM) labeled siRNA nanoparticles fabricated from different formulations for 24 hrs. Cellular internalization of the siRNA nanoparticles was analyzed by fluorescence and confocal microscopes.

Quantitative reverse transcriptase-polymerase chain reaction (RT-PCR) was used for the analysis of the suppression of *BCL2* gene expression in the A549 human lung cancer cells. Gene expression was calculated as the ratio of mean band density of analyzed reverse transcriptase-PCR product to that of the internal standard ($\beta 2$-m).

RESULT AND DISCUSSION

Formulation and characterization of siRNA nanoparticles

In order to develop efficient, multifunctional, and nontoxic siRNA delivery agents for cancer therapy, we prepared and used 5 and 10 nm SPIO nanoparticles for this study. The surface of the as-prepared SPIO nanoparticles was modified with two different polymer layers in tandem. The first layer was introduced by PMAO. Under basic conditions, the anhydride groups of PMAO hydrolyzed to carboxylic groups, which have buffering capabilities at pH 5-6. The carboxylic acid groups make the SPIO nanoparticles hydrophilic and reactive to electrostatically adsorb the second layer of a positively charged PDDA which contains quaternary ammonium functional groups. The advantage of using quaternary over primary amine-containing polymers could be explained by the extremely strong binding and protection of siRNAs. The availability of carboxylate ions in PMAO are thought to accept the protons pumped into endosomes during the pH changes from 7.4 to 5. It results in the swelling and bursting of the endosomes, which would enhance the transport of the siRNA from the endosomes to the cytoplasm ('proton-sponge effect' [5]). Furthermore, when pH changed to ~5-6 in the endosome, the ionized carboxylic acid species (COO-) are transformed to protonated groups to COOH, consequently the PDDA layers are separated from the surface of SPIO. The pH triggered polymer adsorption and desorption from the SPIO surface will greatly facilitate siRNAs dissociation from the gene delivery complexes, since the interaction between PDDA and siRNA is relatively weak compared to the PDDA on the SPIO nanoparticles .

The efficiency of the prepared SPIO nanoparticles to provoke siRNA condensation was studied by ethidium bromide dye displacement assay. As shown in Figure 1, 5 nm SPIO nanoparticles are more efficient in siRNA packaging compared to the larger ones (10 nm). And it is even more efficient than the fifth generation of PPI dendrimer, which has been demonstrated extremely effective in DNA condensation [6].

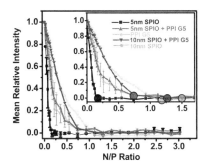

Figure 1. Ethidium bromide dye displacement by PPI G5 dendrimer, 5 and 10 nm SPIO nanoparticles and their mixtures. The highlighted areas on the graphs demonstrate the N/P ratios which correspond to the apparent end of siRNA complexation.

The difference in the condensation strength of 5 nm SPIO nanoparticles and PPI G5 related to the presence of the permanent positive charge represented by the quaternary ammonium of PDDA on the surface of SPIO nanoparticles, which eliminate the charge mobility issue in primary amines at neutral pH. While the decrease in condensing efficiency of larger (10 nm) nanoparticles might be associated with lower positive charge density on their outer surfaces. To integrate targeting groups in the formed siRNA nanoparticles, the condensation efficiency of SPIO nanoparticles mixed with PPI G5 at different ratios has been studied as well. Quantitative analysis of the condensing efficiency of the mixtures reveals that the mixture of 5 nm SPIO nanoparticles and PPI G5 is more efficiently in provoking siRNA condensation than the mixture of 10 nm SPIO and PPI G5.

siRNA complexes formed with above mentioned complexation agents were formulated at N/P ratios, obtained by ethidium bromide displacement assay which represent the apparent end of the complexation. AFM measurments demonstrated (Fig. 2) that SPIO nanoparticles as well as their mixture with PPI G5 were able to effectively package siRNA into discrete particles with an average diameter of ~250 nm and height of ~14 nm. Detailed analysis of the condensates demonstrated that SPIOs were included in the siRNA nanoparticles.

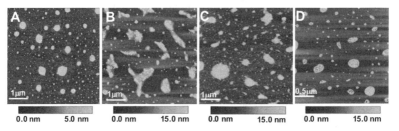

Figure 2. Representative AFM images of condensates formed by siRNA in the presence of (A) 5 nm SPIO; (B) 10 nm SPIO; (C) mixture of 5 nm SPIO with PPI G5 and (D) mixture of 10 nm SPIO with PPI G5 after 30 min of condensation.

In vitro transfection of siRNA nanoparticles

Confocal microscopy was used to evaluate the ability of these multifunctional delivery agents in facilitating internalization of FAM-labeled siRNAs into A549 lung cancer cells (Fig. 3). The ability of the condensing agents to enhance the cellular internalization of the siRNAs declined in the following order: 10 nm SPIO/PPI G5 > 5 nm SPIO/PPI G5 > 5 nm SPIO > 10 nm SPIO. The siRNA complexes formed were distributed in cytoplasm and in the perinuclear region. Since small interfering RNA functions by binding to RNA-induced silencing complex in the cytoplasm, so that the delivery of siRNA to cytosol may have the advantage of avoiding toxicity to the nucleus.

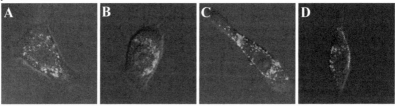

Figure 3. Representative fluorescence confocal microscopic images of cellular uptake of modified FAM-labeled siRNA nanoparticles condensed with (A) 5 nm SPIO; (B) 10 nm SPIO; (C) mixture of 5 nm SPIO with PPI G5, and (D) mixture of 10 nm SPIO with PPI G5.

To investigate the biological activity of SPIO based gene delivery vectors, we carried out the experiment with siRNA, which specifically target the expression of *BCL2* protein. The *BCL2* protein regulates the mitochondria-mediated apoptosis pathway, and various cell death stimuli, including chemotherapeutic agents, activate caspases by this pathway, thereby promoting apoptosis [7]. The results show that treatment with siRNA/SPIO complexes significantly influence the expression of the *BCL2* gene. In contrast, siRNA/PPI G5 complexes demonstrated less suppression efficiency of the targeted gene (Fig. 4). At the studied conditions, 62% and 72% silencing were achieved with 5 nm and 10 nm SPIO nanoparticles, respectively, while in case of PPI G5, silencing was evaluated to be 43%. On the hand, condensing mixtures (SPIO + PPI G5) result in ~80% suppression of the targeted gene suppression (Fig. 4B). Taken together these data show that siRNA complexed with SPIO nanoparticles substantially enhanced their specific activity and in turn resulted in the decrease of the expression of targeted genes.

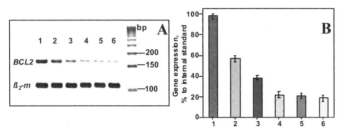

Figure 4. Effect of treatment with (1) no treatment; (2) PPI G5/siRNA; (3) 5 nm SPIO/siRNA; (4) 10 nm SPIO/siRNA; (5) 5 nm SPIO/PPI G5/siRNA; (6) 10 nm SPIO/PPI G5/siRNA on the expression of BCL2 mRNA in A 549 cancer cells.

Sterically stablilze and targeted deliver the siRNA nanoparticles to cancer cells

Another important step in developing multifunctional nanomedicine platforms is to extend the circulation time of the siRNA nanoparticles in blood stream and to specifically deliver the siRNA nanoparticles into cancer cells. To integrate targeting groups to the formed siRNA nanoparticles as well as to increase their sterical stability, siRNA complexes formed by the mixture of SPIO nanoparticles and PPI G5 dendrimer were modified with polyethylene glycol (PEG). The distal end of PEG was modified with targeting agent, such as LHRH as shown in Figure 5.

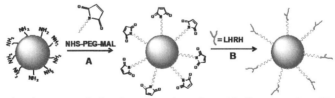

Figure 5. Engineering approach for the preparation of specifically targeted, stable siRNA nanoparticles. (A) PEGylation. (B) Conjugation of targeting peptide to the distal end of the PEG layer.

The selective internalization of the siRNA nanoparticles was demonstrated by incubation of modified siRNA nanoparticles with lung cancer cells A549 and LHRH negative SKOV-3 cancer cells (Fig. 6). The ability of the delivered siRNA to silence their targeted mRNA was demonstrated by RT-PCR experiment (Fig. 6E). The LHRH modified siRNA nanoparticles condensed with the mixture of SPIO nanoparticles and PPI G5 can sufficiently reduce the expression of BCL2 gene.

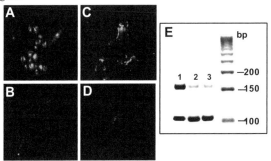

Figure 6. Representative fluorescence microscopic images of cellular uptake of the LHRH-PEG-modified FAM-siRNA nanoparticles condensed with (A, B) 5nm SPIO + PPI G5 and (C, D) 10 nm SPIO + PPI G5 by LHRH positive, human lung carcinoma A549 cells and LHRH negative, ovarian SKOV-3 cancer cells, respectively. Panel E shows the effect of treatment with (1) no treatment, (2) LHRH-PEG-siRNA-SPIO 5 nm-PPI G5 and (3) LHRH-PEG-siRNA-SPIO 10 nm-PPI G5 on the expression of BCL2 mRNA in A 549 cancer cells.

CONCLUSIONS

We envision that the siRNA nanoparticles condensed with modified SPIO can be used as multifunctional siRNA delivery agents for effective cancer therapy, with the ability to real time monitor the therapeutic outcome of combined hyperthermic and siRNA therapy.

ACKNOWLEDGMENTS

The research was supported in part by NIH grants CA100098, CA111766, CA074175 from the National Cancer Institute and Charles & Johanna Bush Biomedical Grants.

REFERENCES

[1] G.R. Devi, *Cancer Gene Ther.* **13,** 819 (2006)
[2] Z. Medarova, W. Pham, C. Farrar, *Nature Medicine* **13,** 372 (2007)
[3] J.L. Zhang, R.S. Srivastava, R.D. Misra, *Langmuir* **11,** 6342 (2007)
[4] S.S. Dharap, T. Minko, *Pharm, Res.* **20,** 889 (2003)
[5] Y. Koyama, T. Ito, H. Matsumoto, et al., *J. Biomater. Sci. Polymer. Edn.* **14** 515 (2003)
[6] A.M. Chen, L.M. Santhakumaran, S.K. Nair, et al. *Nanotechnology* **17** 5449 (2006)
[7] S.S. Dharap, P. Chandna, Y. Wang, et al., *JPET* **316** 992 (2006)

Mater. Res. Soc. Symp. Proc. Vol. 1140 © 2009 Materials Research Society 1140-HH12-01

Functionalized Magnetic Nanoparticles for Selective Targeting of Cells

Tremel, Wolfgang; Shukoor, Mohammed; Natalio, Filipe; Tahir, Muhammad; Wiens, Matthias; Schladt, Thomas; Barz, Matthias; Theato, Patrick; Schröder, Heinz; Müller, Werner

Universität Mainz - Institut für Anorganische Chemie und Analytische Chemie, Mainz, Germany

ABSTRACT

MnO nanoparticles were conjugated to single stranded DNA (ssDNA), Cytosin-phosphatidyl-Guanosin oligonucleotide (CpG ODN) to detect and activate Toll-like (TLR9) receptors in cells and to follow nanoparticle cellular trafficking by different means of imaging while at the same time serving as a drug carrier system. By virtue of their magnetic properties these nanoparticles may serve as vehicles for the transport of target molecules into cells, while the fluorescent target ligand allows optical detection simultaneously.

INTRODUCTION

One of the difficult aspects of developing an *in vivo* approach to cancer treatment is the specific targeting of cancer cells. One strategy is the use of therapeutic nucleotides. Microbial pathogens that penetrate epithelial barriers and invade tissues are usually encountered by three types of sentinel immune cells: tissue macrophages, mast cells and immature dendritic cells. These sentinels must be able to distinguish between fragments of apoptotic cells generated during normal tissue turnover and particles that are indicative of microbial assaults and infections. The molecules responsible for making this pivotal distinction belong to the family of pattern recognition receptors (PRRs), of which Toll-like receptors (TLRs) are best characterized [1]. TLRs recognize highly conserved microbial structures that were termed pathogen-associated molecular patterns (PAMPs). Stimulation of macrophages or mast cells through their TLRs leads to the synthesis and secretion of proinflammatory cytokines and lipid mediators, thereby initiating an inflammatory response that recruits both soluble immune components and immune cells from the blood [2]. On the other hand, TLR stimulation of dendritic cells induces the initiation of an adaptive immune response [3].

One of the exciting new research subjects involving magnetic nanoparticles is their application in biological systems, including targeted drug delivery, magnetic resonance imaging (MRI), biosensors and magnetic hyperthermia therapy. Nanoparticles are attractive probe candidates because of their (i) small size (1-50 nm) and correspondingly large surface-to-volume ratio, (ii) chemically tailorable physical properties which directly relate to size, composition, and shape, (iii) unusual target binding properties, and (iv) overall structural robustness. The size of a nanomaterial can be an advantage over a bulk structure, because a target binding event involving the nanomaterial can have a significant effect on its physical and chemical properties, thereby providing a mode of signal transduction not necessarily available with a bulk structure made of the same material.

We have designed a pathogen-mimicking metal oxide nanoparticle with the ability to enter cancer cells and selectively target and activate the TLR9 pathway, in addition to optical and

MR imaging capabilities. The immobilization of ssDNA, CpG-ODN on MnO nanoparticles was performed via a multifunctional polymer used for the nanoparticle surface modification. The polymer coating not only affords a protective organic biocompatible shell but also provides an efficient and convenient means for loading immunostimulatory oligonucleotides.

EXPERIMENT

Manganese oxide nanoparticles were synthesized according to ref. [4]. Phase identification of the naked manganese oxide nanoparticles was carried out using transmission electron microscopy (TEM) (on a Philips 420 instrument with an acceleration voltage of 120 kV or on a Philips TECNAI F30 electron microscope, field-emission gun, 300 kV extraction voltage) and X-ray powder diffraction (Siemens D8 powder diffractometer). The magnetic susceptibilities of samples were measured with a SQUID magnetometer (Quantum Design).

The particles were functionalized using a multidentate functional copolymer [5] carrying catecholate groups as surface binding ligands for the manganese oxide nanoparticles, a fluorescent dye for optical detection and free amino groups for the attachment of the target ligands. An aliquot MnO was treated with an appropriate amount of the reactive polymer dissolved in N,N-dimethylformamide (DMF). To remove unbound polymer the coated magnetic particles in the solution were extracted by a magnetic particle concentrator (Dynal MPC1-50, Dynal Biotech, France) at room temperature. The isolated magnetic nanoparticles were washed with DMF ensuring the removal of unreacted polymer and subsequently dispersed in methyl imidazole buffer (MeIm, 0.1 M, pH 7.5). A portion of the washed magnetic particles was freeze-dried for subsequent characterization. The average crystallite size of the particles with and without the functional polymer coating was estimated using TEM.

The presence of primary amine groups on the surface-bound polymer ligand permits the attachment of reactive N-hydroxy-succinimide ester of porphyrin through peptide chemistry (experimental details are supplied in ESI). dsRNA CpG was obtained commercially (Sigma) and prepared to a final concentration of 2 mg/ml. dsRNA CpG contains phosphate groups at its 5' end which makes it susceptible to functionalization in the presence of a primary amine making use of phosphoramidate chemistry [6].

RESULTS AND DISCUSSION

XRD measurements of the as prepared manganese oxide show the formation of the MnO phase. This is further confirmed by TEM analysis. The MnO particles were prepared with sizes ranging from 6-25 nm, Figure 1 shows a sample of MnO nanoparticles with diameters of 7 nm. The hysteresis loops recorded at low temperature (10 K) showed large coercitivities and large remanences typical for ferrimagnetic materials. Particles exhibit paramagnetism at room temperature. It has been recently reported that very small MnO nanoparticles (5–10 nm in diameter) show weak ferromagnetic behavior at low temperatures [4a], although bulk MnO shows antiferromagnetic behavior with T_N =125 K. The observed weak ferromagnetism was ascribed to the presence of noncompensated surface spins on the antiferromagnetic core of the MnO nanoparticle [4a].

In this contribution we introduce a novel multifunctional polymeric ligand (Figure 2) to immobilize CpG ODN onto MnO nanoparticles that curbs the multistep tailoring of nanoparticle functionalization. The multifunctional polymeric ligand combines three features: (i) an anchor group based on dopamine which is capable of binding to many metal oxides, [7] (ii) a fluorescent dye (as a marker) and (iii) a reactive functional group which allows binding of various biomolecules onto inorganic nanoparticles. CpG ODN contains a 5' end phosphate group which makes it amenable to bind to amine moieties by making use of phosphoramidate chemistry [6]. The biological activity of the CpG ODN coated magnetic nanoparticles was demonstrated on kidney cancer cells Caki-1 (human renal cell line). The specific binding of the nanoparticle CpG ODN complex to the cell receptors was proven experimentally as sketched in Figure 3.

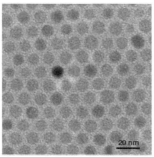

Figure 1. Transmission electron microscopic (TEM) image of manganese oxide nanoparticles.

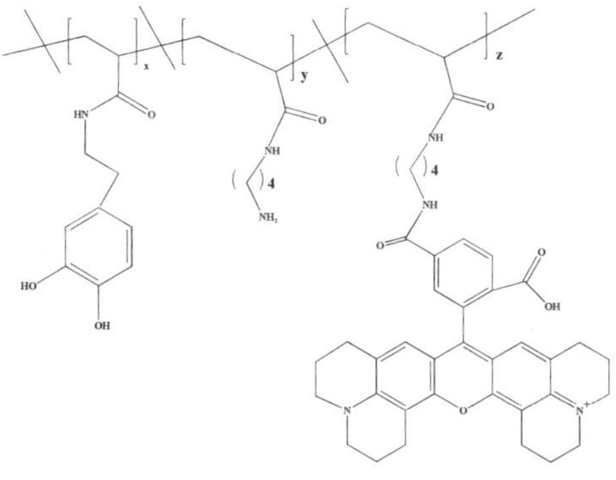

Figure 2. Multifunctional copolymer containing 3-hydroxytyramine (dopamine) as an anchor group for the binding of metal oxides, rhodamine, as a fluorophore, and a free amine group for the conjugation of ssDNA, CpG ODN.

The presence of TLR9 in Caki-1 cells was determined using NaDoSO$_4$. The correspondent protein identification and localization of receptors was achieved through western blotting and immunocytochemistry, respectively (Figure 3). TLR9 is known to be localized intracellularly [23] and it comprises of a ligand-binding ectodomain and a cytoplasmatic portion which belongs to the Tol-IL-1 receptor (TIR) family of signalling domains [8]. The Caki-1 cells were lysed and the supernatant was analyzed using a standard gel electrophoresis procedure. A 12% NaDoSO$_4$ gel was used to separate the polypeptides, the proteins were transferred to a membrane and treated with monoclonal anti-TLR9 antibodies (1:500 dilution). The immunocomplexes were visualized after the incubation with anti-mouse IgG (*Fab*-specific) alkaline phosphatase produced in goat (1:1000) using a color developing system (NBT/BCIP). The western blotting analysis shows a band of approximately 130 kDa, (Figure 3A), which is in agreement with the glycosylated molecular weight of TLR9 [9]. For the localization of TLR9 in the Caki-1 cells, an immunodetection technique was used (Figure 3B). Following methanol fixation, the cells were rehydrated and incubated with TLR9 mouse monoclonal antibody raised against human TLR9 (hTLR9), which in turn were conjugated with Cy-3-labeled anti-mouse IgG secondary antibody (Figure 5B). Subsequently, the cells were analyzed using optical light microscopy, with a reflected light fluorescence attachment at the emission wavelength of 456 (filter U) and 565 nm (filter G) to visualize the DAPI and the Cy-3 flourophore, respectively (Figure 3B (b)). The permeabilization step using methanol was necessary to stain the intracellularly distributed TLR9. A red fluorescence signal was obtained due to Cy-3 -labelled secondary antibody, proving the presence of TLR9 in Caki-1 cells, in accordance with the NaDoSO$_4$/WB results. Control was performed by replacing the primary antibody with a blocking solution where no fluorescent signal was observed (Figure 3B (a)). We have shown that the TLR9 have a molecular weight of 130 kDa, owing to its glycosylated state, and is located intracellularly in Caki-1 cells suggesting its localization in the endoplasmatic reticulum (ER) as described for ssDNA-CpG unstimulated cells [10]. The internal localization of the TLR9 regulates the access to its ligands due to biological reasons [11].

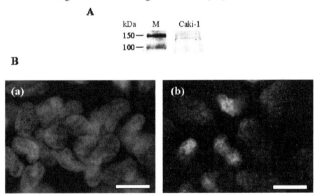

Figure 3. Identification of Toll-like receptor 9 in Caki-1 cells by Western blotting (**A**). Cell lysate was prepared and separated by gel electrophoresis. The western blotting experiments the membrane was reacted with monoclonal antibodies raised against human TLR9. A clear band with a molecular weight of approximately 130 kDa can be identified. M: a size marker was run in parallel. Epiflourescent images of immunostaining of TLR9 in Caki-1 cells (**B**). TLR9 monoclonal antibody was used as well as the correspondent Cy3-fluorophore conjugated secondary antibody (red). The red fluorescence signal is clearly visualized in the Caki-1 cells. Controls were performed using blocking solution PBS/FCS (10%v/v) as replacement of the

primary antibody. Nuclei were visualized by staining with 4,6-diamino-2-phenylindole (DAPI) (blue signal) (scale bar: 20 μm).

FITC-CpG ODN conjugated–polymer functionalized MnO nanoparticles were incubated with Caki-1 cells (37 °C, 5% CO_2) for 24h to target and promote TLR9 mediated cascade activation. Figure 4 shows epifluorescent microscope images representative of the extent of DNA conjugated nanoparticles in the cytosol. Following incubation, the cells were analyzed with a reflected light fluorescence attachment at different emission wavelengths to visualize DAPI staining (blue fluorescent signal), polymer functionalized MnO nanoparticles (red fluorescent signal) and FITC-CpG ODN (green fluorescent signal). The colocalization of green/red fluorescence signals in Figure 4 results in an overlap of fluorescent signals supporting polymer functionalized MnO nanoparticles as efficient carriers of ssDNA for the Caki-1 cells. After 24h incubation the co-localized image shows an increased accumulation of FITC-CpG ODN loaded polymer functionalized nanoparticles into cellular compartments, which can now be attributed to lysosomes [10a].

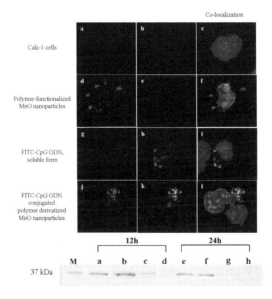

Figure 4. Top: Cellular uptake of MnO nanoparticles functionalized with rhodamine-polymer (red fluorescence) and coupled with FITC-CpG ODN (green fluorescence) at 24h incubation. The MnO-polymer-ssDNA nanoparticles were found to be localized in the cytosol, forming small vesicles (lysosomes). The nuclei were visualized by DAPI (blue). Bottom: The specific binding between FITC-CpG ODN-polymer functionalized MnO nanoparticles and TLR9 in caki 1 cells activates signalling pathways resulting in ubiquination of IkBα (eg. an immunomodulated response), as confirmed from the western blotting analysis.

Interestingly, neither phosphodiester nor peptide bonds in the ssDNA CpG ODN coupled functionalized nanoparticles have been disrupted by lysosomal hydrolases. No green fluorescent signal randomly dispersed in the cytosol was visualized, demonstrating, polymer derivatized MnO nanoparticles to be an efficient cargo delivery system. The unspecific uptake was taken into consideration by using the polymer functionalized nanoparticles.

CONCLUSIONS

In summary, we have designed the first pathogen-mimicking metal oxide nanoparticles with the ability to enter cancer cells and selectively target and activate the TLR9 pathway, in addition to optical and MR imaging capabilities. The multifunctional polymer used for the nanoparticle surface modification not only affords a protective organic biocompatible shell but also provides an efficient and convenient means for loading the immunostimulatory oligonucleotides. The development of versatile trimodal nanoparticles would allow a common user to follow nanoparticle cellular trafficking by different means of imaging and simultaneous use as drug effective carrier system. The generality of this approach should allow the design of nano-based therapeutics that can specifically target a wide range of diseased cells.

ACKNOWLEDGMENTS

We are grateful to the Materials Science Center (MWFZ) of the University of Mainz for financial support.

REFERENCES

1. E. G. Pamer, *Nat. Immunol.* **8**, 1173 (2007).
2. K. Takeda, T. Kaisho, S. Akira, *Annu. Rev. Immunol.* **21**, 335 (2003).
3. F. Re, J. L. Strominger, *Immunobiology* **209**, 191 (2004).
4. (a) G. H. Lee, S. H. Huh, J.W. Jeong, B. J. Choi, S. H. Kim, H.-C. Ri, *J. Am. Chem. Soc.* **124**, 12094 (2004). (b) M.Yin, S. O´Brien, *J. Am. Chem. Soc.* 125, 10180 (2003). (c) W. S. Seo, H. H. Jo, K. Lee, B. Kim, S. J. Oh, J. T. Park, *Angew. Chem. Int. Ed.* **43**, 1115 (2004).
5. M. N. Tahir, M. Eberhardt, P. Theato, S. Faiß, A. Janshoff, T. Gorelik, U. Kolb, W. Tremel, *Angew. Chem. Int. Ed.* **45,** 908 (2006).
6. B. C. F. Chu, G. M. Wahl, L. E. Orgel, *Nucl. Acids Res.* **11**, 6513 (1983).
7. M. N. Tahir, M. Eberhardt, P. Theato, S. Faiss, A. Janshoff, T. Gorelik, U. Kolb, W. Tremel, *Angew. Chem. Int. Ed.* **45,** 908 (2006).
8. C. A. Leifer, M. N. Kennedy, A. Mazzoni, C. W. Lee, M. J. Kruhlak, D. M. Segal, *J. Immun.* **173**, 1179 (2004).
9. G. M. Barton, J. C. Kagan, R. Medzhitov, *Nat. Immun.* 7, 49 (2006).
10. (a) P. Ahmad-Nejad, H. Häcker, M. Rutz, S. Bauer, R. M. Vabulas, H. Wagner, *Eur. J. Immunol.* **32**, 1958 (2002).
11. C.A. Leifer, J. C. Brooks, K. Hoelzer, J. Lopez, M. N. Kennedy, A. Mazzoni, D. M. Segal, *J. Biol. Chem.* **281**, 35585 (2006).

Mater. Res. Soc. Symp. Proc. Vol. 1140 © 2009 Materials Research Society   1140-HH13-14

Delivery of Doxorubicin to Solid Tumors Using Thermosensitive Liposomes

Elizaveta V. Tazina, Alevtina P. Polozkova, Elena V. Ignatieva, Olga L. Orlova,

Valeria V. Mescherikova, Adolf A. Wainson, Natalia A. Oborotova,

Anatoliy Yu. Baryshnikov

Research Institute of Experimental Diagnostics and Therapy of Tumors,

N.N. Blokhin Russian Cancer Research Center of RAMS,

24 Kashirskoye shosse, 115478, Moscow, Russian Federation

ABSTRACT

Liposomal drugs currently in clinical use are characterized primarily by their decreased side effects rather than improved therapeutic potency. Significant improvements in the efficacy of liposomal drug therapy may be obtained using thermosensitive liposomes (TSL) in combination with local hyperthermia (HT). The purpose of present work was preparation of TSL loaded with doxorubicin (Dox) and investigation of their effect on B-16 mouse melanoma and Ehrlich (line ELD) carcinoma in combination with HT. TSL were prepared using 1,2-dipalmitoyl-sn-glycero-3-phosphocholine, 1,2-distearoyl-sn-glycero-3-phosphocholine, PEGylated 1,2-distearoyl-sn-glycero-3-phosphoethanolamine, cholesterol and α-tocopherol acetate in molar ratios of 9:1:0.02:0.2:0.2 (composition 1); 9:1:0.1:0.5 (composition 2); 9:1:0.2:0.75 (composition 3); 9:1:0.3:1 (composition 4); 9:1:0.4:1.2 (composition 5). Dox was loaded into TSL by ammonium ion gradient. Efficacy of Dox encapsulation in TSL of compositions 4 and 5 was 60 % (diameter of vesicles was 175 ± 10 nm). TSL of compositions 2 and 3 encapsulated 88 % and 86 % of Dox, respectively (diameter of vesicles was 160 ± 10 nm). TSL of composition 1 trapped 88-94 % of Dox (diameter of vesicles was 175 ± 15 nm). Anticancer efficacy of Dox-TSL (composition 1) and free Dox was compared in biological experiments. The doubling time of B-16 melanoma was 9 days after heating on a background of Dox injection at dose of 9 mg/kg, while heating of tumors after injection of Dox-TSL at doses of 4.5 and 9 mg/kg increased tumor doubling time up to 12 and 16 days, respectively. The doubling time of Ehrlich carcinoma increased from 3 days in the control group up to 14 days for the group of mice administered 9 mg/kg of Dox-TSL followed by HT in 15-20 min. Thus, Dox-TSL in combination with HT has shown more efficiency than free Dox in suppression of tumor growth.

INTRODUCTION

Liposomes are lipid vesicles with an aqueous interior size between 50 nm and 200 nm in diameter. They form spontaneously when certain lipids are dispersed in aqueous medium. To circumvent the rapid uptake of liposomes in liver and spleen after intravenous application, the surface of liposomes has been modified with poly(ethylene glycol) (PEG) [1]. These PEG-liposomes show extended plasma half lives of several hours in humans enabling passive accumulation of liposomes in solid tumors, where the microvessels are leakier (Enhanced Permeability and Retention (EPR) effect) [2].

Liposomes are now recognized for their ability to increase the therapeutic activity and/or reduce the toxicity of selected encapsulated chemotherapeutic agents. Side effects of cytotoxic drugs like doxorubicin (Dox) are significantly reduced by liposomal encapsulation. This has been demonstrated in various clinical studies using PEGylated-liposomal Dox (Caelyx®, Doxil®) and non-PEGylated Dox (Myocet™) with regard to cardiotoxicity [3, 4]. However, compared to free drug, no improved effectiveness of these clinically approved liposomal formulations has been reported [5]. In some cases, the reason for the lack of improvement in therapeutic potency is the low drug release rate from traditional liposomes [6]. It has therefore been proposed that significant improvements in the efficacy of liposomal drug therapies may be obtained by addition of a "trigger" to the liposomal membrane that facilitates the release of liposomal contents after liposomes have accumulated at the target site [7]. One method of triggering drug release is the use of thermosensitive liposomes (TSL) in combination with local tumor hyperthermia (HT) [8].

Triggered drug release from TSL was first described by Yatvin et al. in 1978 [9]. TSL were developed over the last decades [10-14]. TSL lack or contain reduced amounts of cholesterol and release their contents at a specific formulation-dependent gel to liquid-crystalline phase-transition temperature (T_m). At this temperature, the membrane's permeability increases by several orders of magnitude [8]. At T_m lipids change from the "solid" gel to the "fluid" liquid crystalline state. A conformational change of C–C single bonds in the alkyl chains of the lipids leads to an increase in the total volume occupied by the hydrocarbon chains in the membrane, and therefore increases the permeability of the bilayer membrane. TSL mainly composed of 1,2-dipalmitoyl-sn-glycero-3-phosphocholine (DPPC; T_m = 41.4 °C [15]) and 1,2-distearoyl-sn-glycero-3-phosphocholine (DSPC; T_m = 54.9 °C [15]), release the drug at temperature of about 43 °C.

The purpose of present research was preparation of TSL with encapsulated Dox and assessment of biological activity *in vitro* and *in vivo*.

EXPERIMENTAL DETAILS

Preparation of TSL loaded with Dox

TSL were prepared by reverse evaporation method using 1,2-dipalmitoyl-sn-glycero-3-phosphocholine (DPPC), 1,2-distearoyl-sn-glycero-3-phosphocholine (DSPC) (Lipoid), 1,2-distearoyl-sn-glycero-3-phosphoethanolamine-N-[methoxy(polyethylene glycol)-2000] (DSPE-PEG 2000 ammonium salt), cholesterol (Avanti Polar Lipids) and α-tocopherol acetate in molar ratios of 9:1:0.02:0.2:0.2 (composition 1); 9:1:0.1:0.5 (composition 2); 9:1:0.2:0.75 (composition 3); 9:1:0.3:1 (composition 4); 9:1:0.4:1.2 (composition 5). The extrusion was performed through 200 nm polycarbonate membranes (Whatman) at 50 °C using Avanti Mini-Extruder. Dox was loaded into TSL by ammonium ion gradient for 1 h at 50 °C. Drug-to-lipid ratio was 0.13:1 (w/w). The ammonium ion gradient was formed by dilution of the liposomal dispersion 20 times in 10 mM HEPES (pH 8.4). TSL loaded with Dox (Dox-TSL) were lyophilized for better stabilization with 4 % sucrose added as a cryoprotectant. Vesicle size was evaluated by dynamic laser light-scattering measurements using Nicomp-380 Submicron Particle Sizer. Dox-TSL were separated from untrapped Dox by gel filtration on column C 10/20 (Amersham Biosciences) with Sephadex G-50 sorbent and 0.15 M sodium chloride solution as an eluent. The concentration of Dox in TSL was determined spectrophotometrically at wavelength of 252 nm.

Effect of Dox-TSL on B-16 melanoma cells *in vitro*

Dox-TSL and free Dox *in vitro* were administered to B-16 melanoma cell suspension at concentrations corresponding to 3-15 µg of Dox per ml. Cell suspension with drugs was incubated during 1 h at 37 °C or 42.5 °C. The suppression of cell growth and cell survival (clonogenicity) were used as the end points.

Effect of Dox-TSL on transplantable tumors under HT

Dox-TSL or free Dox were administered to C57BL/6j mice with B-16 melanoma and (CBA x C57BL) mice with Ehrlich (line ELD) carcinoma on the 8-th day after transplantation. Tumors were transplanted in the shank muscle of mice. Dox-TSL or free Dox were injected in retroorbital sinus of animals at doses of 9 and 4.5 mg Dox per kg body weight. HT treatment of shank with transplanted tumor was performed at 43 °C for 30 min using water bath. Each treated group included 6 animals. There were slight differences in the arrangement of B-16 melanoma and Ehrlich carcinoma experiments.

The animals bearing B-16 melanoma were divided into 8 groups:
1 – Control untreated group; 2 – Animals administered 4.5 mg/kg of Dox-TSL; 3 – Animals administered 9 mg/kg of Dox-TSL; 4 – Animals administered 9 mg/kg of Dox; 5 – Animals treated with HT; 6 – Animals administered 4.5 mg/kg of Dox-TSL + HT; 7 – Animals administered 9 mg/kg of Dox-TSL + HT; 8 – Animals administered 9 mg/kg of Dox + HT.

The animals with Ehrlich carcinoma received the following treatment:
1 – Control untreated group; 2 – Animals administered 9 mg/kg of Dox-TSL; 3 – Animals treated with HT; 4 – Animals administered 9 mg/kg of Dox-TSL followed by HT in 10-12 min; 5 – Animals administered 9 mg/kg of Dox-TSL followed by HT in 15-20 min; 6 – Animals administered 9 mg/kg of Dox-TSL followed by HT in 35-40 min; 7 – Animals administered 9 mg/kg of Dox followed by HT in 15-20 min.

The tumors were measured three times a week and mean tumor volume in the group was calculated. Dynamics of tumor regression and regrowth was assayed as ratio (Vt/Vo) of the mean tumor volume at the day of measurement (Vt) to the mean tumor volume at the day of experiment (Vo). Animals were followed till their death.

RESULTS

Efficacy of Dox encapsulation in TSL of compositions 4 and 5 was 60 % (diameter of vesicles was 175 ± 10 nm). TSL of compositions 2 and 3 encapsulated 88 % and 86 % of Dox, respectively (diameter of vesicles was 160 ± 10 nm). TSL of composition 1 trapped 88-94 % of Dox (diameter of vesicles was 175 ± 15 nm). Sizes distribution for Dox-TSL of composition 1 is shown in figure 1. The composition 1 has been chosen for preparation of the drug "Lyophilized thermosensitive liposomal Doxorubicin for injection 0.7 mg," which has been assessed in biological experiments *in vitro* and *in vivo*.

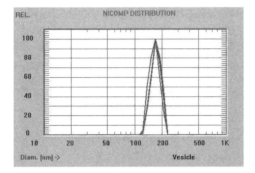

Figure 1. Sizes distribution of Dox-TSL determined by Nicomp-380 Submicron Particle Sizer.

In the experiments *in vitro* the heating of Dox-TSL has resulted in more than 10-fold decrease of cells growth rate. In the presence of free Dox cells growth rate decreased two times at 42.5 °C compared with cells growth rate at 37 °C (figure 2A). Cells death increased approximately fifteen times when cell suspension was incubated with Dox-TSL at elevated temperature instead of 37 °C (figure 2B).

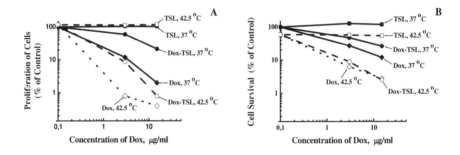

Figure 2. Proliferation of B-16 melanoma cells (**A**) and survival of cells (**B**) after incubation with free Dox or Dox-TSL at 37 °C and 42.5 °C.

The doubling time of tumors was 9 days after heating on a background of Dox injection at dose of 9 mg/kg, while heating of tumors after injection of Dox-TSL at doses of 4.5 and 9 mg/kg increased tumor doubling time up to 12 and 16 days, respectively (figure 3A).

By the 34th day after treatment three mice have survived in the group administered 9 mg/kg of Dox-TSL and treated with HT. In the group of animals administered 4.5 mg/kg of Dox-TSL + HT only one mouse has survived by this time. The 50 % survival has been observed during 34 and 29 days in these groups, respectively (figure 3B). In the group of mice administered 9 mg/kg of free Dox + HT the 50 % survival has been observed during 21 days.

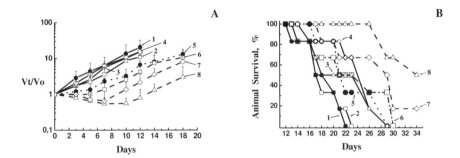

Figure 3. Dynamics of B-16 melanoma growth (**A**) and survival of animals with B-16 melanoma (**B**) after HT treatment on a background of Dox and Dox-TSL administration.
1 – Control; 2 – Dox (9 mg/kg); 3 – Dox-TSL (9 mg/kg);
4 – Dox-TSL (4.5 mg/kg); 5 – HT (43 °C, 30 min);
6 – Dox (9 mg/kg) + HT; 7 – Dox-TSL (4.5 mg/kg) + HT;
8 – Dox-TSL (9 mg/kg) + HT.

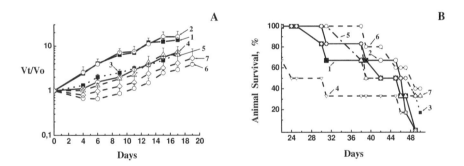

Figure 4. Dynamics of Ehrlich carcinoma growth (**A**) and survival of animals with Ehrlich carcinoma (**B**) after HT (43 °C, 30 min) treatment on a background of Dox and Dox-TSL administration at dose of 9 mg/kg.
1 – Control; 2 – Dox-TSL; 3 – HT;
4 – Dox + HT in 15-20 min;
5 – Dox-TSL + HT in 10-12 min;
6 – Dox-TSL + HT in 15-20 min;
7 – Dox-TSL + HT in 35-40 min.

The slowest growth rate of Ehrlich carcinoma has been observed in the group of mice administered 9 mg/kg of Dox-TSL followed by HT in 15-20 min. The doubling time of tumors increased from 3 days in the control group up to 14 days for this group (figure 4A). Tumor growth inhibition after Dox-TSL injection followed by HT in 10-12 min and 35-40 min was less significant.

The highest survival rate was noticed in the group of animals administered 9 mg/kg of Dox-TSL followed by HT in 15-20 min. The 100 % survival has been observed during 38 days for this group (figure 4B). By the 22nd day after treatment two mice have died in the group administered 9 mg/kg of free Dox followed by HT in 15-20 min. All mice have died in the control group by the 50th day of experiment, while two mice have survived by this day in the group treated with HT in 15-20 min after Dox-TSL administration.

CONCLUSIONS

The method of TSL preparation with Dox encapsulation by ammonium ion gradient allows producing the drug with high trapping efficacy of Dox (88-94 %). Both *in vitro* and *in vivo* "Lyophilized thermosensitive liposomal Doxorubicin for injection 0.7 mg" in combination with HT has shown less toxicity and more efficiency compared with free Dox.

REFERENCES

1. T.M. Allen, C. Hansen, F. Martin, C. Redemann, A. Yau-Young, *Biochim. Biophys. Acta* **1066**, 29-36 (1991).
2. D.W. Northfelt, F.J. Martin, P. Working, P.A. Volberding, J. Russell, M. Newman, M.A. Amantea, L.D. Kaplan, *J. Clin. Pharmacol.* **36**, 55-63 (1996).
3. M.E. O'Brien, N. Wigler, M. Inbar, R. Rosso, E. Grischke, A. Santoro, R. Catane, D.G. Kieback, P. Tomczak, S.P. Ackland, F. Orlandi, L. Mellars, L. Alland, C. Tendler, *Ann. Oncol.* **15**, 440-449 (2004).
4. G. Batist, J. Barton, P. Chaikin, C. Swenson, L. Welles, *Expert Opin. Pharmacother.* **3**, 1739-1751 (2002).
5. I. Judson, J.A. Radford, M. Harris, J.Y. Blay, Q. van Hoesel, A. le Cesne, A.T. van Oosterom, M.J. Clemons, C. Kamby, C. Hermans, J. Whittaker, E. Donato di Paola, J. Verweij, S. Nielsen, *Eur. J. Cancer* **37**, 870-877 (2001).
6. S. Bandak, D. Goren, A. Horowitz, D. Tzemach, A. Gabizon, *Anticancer Drugs* **10**, 911-920 (1999).
7. L.D. Mayer, *Cancer Metastasis Rev.* **17**, 211-218 (1998).
8. G. Kong, M.W. Dewhirst, *Int. J. Hypertherm.* **15**, 345-370 (1999).
9. M.B. Yatvin, J.N. Weinstein, W.H. Dennis, R. Blumenthal, *Science* **202**, 1290-1293 (1978).
10. T.P. Chelvi, R. Ralhan, *Int. J. Hypertherm.* **11**, 685-695 (1995).
11. M.H. Gaber, K. Hong, S.K. Huang, D. Papahadjopoulos, *Pharm. Res.* **12**, 1407-1416 (1995).
12. D. Needham, M.W. Dewhirst, *Adv. Drug Delivery Rev.* **53**, 285-305 (2001).
13. L.H. Lindner, M.E. Eichhorn, H. Eibl, N. Teichert, M. Schmitt-Sody, R.D. Issels, M. Dellian, *Clin. Cancer Res.* **10**, 2168-2178 (2004).
14. J.K. Mills, D. Needham, *Biochim. Biophys. Acta* **1716**, 77-96 (2005).
15. S. Mabrey, J.M. Sturtevant, *Proc. Natl. Acad. Sci. U.S.A.* **73**, 3862-3866 (1976).

Mater. Res. Soc. Symp. Proc. Vol. 1140 © 2009 Materials Research Society   1140-HH06-20-DD03-20

Iron Oxide Magnetic Nanoparticles for Treating Bone Diseases

Nhiem Tran[1], Rajesh Pareta[2], and Thomas J. Webster[3]

[1] Department of Physics, Brown University, Providence, RI, 02912. USA

[2] Divisions of Engineering, Brown University, Providence, RI, 02912. USA

[3] Divisions of Engineering and Department of Orthopedic, Brown University, Providence, RI 02912. USA

ABSTRACT

Magnetic nanoparticles have been widely used in biomedical research [1]. Our research goal is to treat bone diseases (such as osteoporosis) by using surface modified magnetic nanoparticles. Magnetite (Fe_3O_4) and maghemite (Fe_2O_3) were synthesized and coated with calcium phosphate (CaP). The resulting nanoparticles were treated thermally to change the crystalline properties of CaP. Nanoparticles were characterized via transmission electron microscopy (TEM) and vibrating sample magnetometry (VSM). Osteoblast (OB) proliferation experiments were conducted after 1, 3 and 5 days in the presence of iron oxide nanoparticles alone, CaP coated iron oxide nanoparticles. OB proliferation experiment were also conducted after 1,3 and 5 days in the presence of various concentrations of CaP coated nanoparticles to show a concentration dependent trend on OB density.

INTRODUCTION

Finding a more effective way to treat bone disease, such as osteoporosis, is of great concern to medical field. Since current drugs used to treat this disease are often times not directed to the diseased site, drug effectiveness is reduced. According to previous research [2], it is believed that through the use of magnetic nanoparticles, an optimal drug delivery system can be developed by using an external magnetic field to direct such nanoparticles to the site of bone disease to immediately increase bone density (Figure 1). The objective of the present study was to determine the influence of magnetic nanoparticles on osteoblast (OB – bone forming cells) functions. Evidence was provided that OB density increased in the presence of various calcium phosphate (CaP) coated magnetic nanoparticles compared to without nanoparticles. Nanoparticle concentration dependence on OB density was also demonstrated.

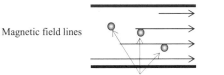

Magnetic field lines

Drug coated magnetic nanoparticles   Tissue

Figure 1: A simplified design for a magnetic drug delivery system in which magnetic nanoparticles coated with drug are directed to bone tissue by an external magnetic field.

EXPERIMENT

The magnetite nanoparticles were prepared by a wet chemical method as previously described [3]. Iron (II) chloride and iron (III) chloride with a molar ratio of 1:2 were dissolved in deoxygenated water in the presence of HCl. The resulting solution was added drop wise to a NaOH solution under vigorous stirring and nitrogen flow. A black precipitate of magnetite (Fe_3O_4) was produced. The precipitate was centrifuged and washed 3 times and redispersed in deoxygenated water. Maghemite nanoparticles were obtained from magnetite by aeration in boiling water at low pH. The final brown-red solution (γ -Fe_2O_3) was centrifuged, washed 3 times and dispersed in deionized water.

The particles were further coated with calcium phosphate (CaP: the main inorganic component of bone) to tailor them to treat osteoporosis. A well-established wet chemistry process was adopted to synthesize CaP [4]. Briefly, 1M calcium nitrate solution (Fisher Scientific, USA) containing iron oxide nanoparticles was added drop wise into a 0.6M potassium phosphate (Sigma) solution under vigorous stirring. The CaP precipitates were washed 3 times with ethanol and deionized water. The precipitation was later treated hydrothermally at 70°C for 20 hours to obtain nanocrystalline CaP. On the other hand, less crystallized CaP was obtained without hydrothermal treatment. The precipitation was directly dried in a vacuum oven at 40°C for 20 hours. Nanoparticles were then dispersed in deionized water. To reduce nanoparticle agglomeration, a common problem encountered when using nanoparticles that decrease their effectiveness, these particle solutions were added with surfactants citric acid (CA), bovine serum albumin (BSA) and dextrant.

To characterize the size of the nanoparticles, a droplet of nanoparticles was placed on a TEM copper grid and allowed to dry. The imaging was carried out on a Phillips EM420 TEM.

The magnetic properties (hysteresis loop) of the dried nanoparticles were obtained using vibrating sample magnometry (VSM, LakeShore 7400) at room temperature. Briefly, magnetic nanoparticle solutions were centrifuged and supernatant decanted. These nanoparticles were then allowed to dry at room temperature for several days. Dried powders were weighed and read by VSM.

Osteoblast proliferation tests were conducted at 1, 3, and 5 days. Cells were grown in Dulbecco's Modified Eagle Medium (DMEM, Gibco) supplemented with 10% fetal bovine serum (FBS, Hyclone, UT) and 1% penicillin/streptomycin (P/S, Hyclone, UT) in a 96 well plate at a volume of 200 µL per well and seeding density of 3500 cells/cm^2 in the presence of iron oxide nanoparticles, CaP coated magnetic nanoparticles, and a control with no nanoparticles. Background solutions containing no cells were also prepared using DMEM supplemented with the same concentrations of nanoparticles as those exposed to cells. Each solution was replicated three times. Nanoparticles were sonicated and vortex mixed before being added at a volume of 20 µl and a concentration of 100 µg/ml to each well. Cells were incubated under standard conditions (37°C, humidified, 5% CO_2,/95% air environment) for 1, 3 and 5 days before the use of the CellTiter96 (Promega) assay. Cell solutions were incubated for two more hours and then analyzed using a microplate reader (SpectraMax300, Molecular Devices) at 490nm for absorbance. A standard curve was also set up to correlate absorbance and cell number.

Osteoblast densities were also measured after 1, 3, and 5 days in the presence of CaP coated Fe_3O_4 magnetic nanoparticles added at various concentrations. For this experiment, Fe_3O_4 magnetic nanoparticles were coated with CaP in the presence of two different surfactants, citric acid (CA) and bovine serum albumin (BSA). The original nanoparticle solutions had concentrations of 1 mg/ml (for the solution that had CA as a surfactant) and 10 mg/ml (for the solution that had BSA as a surfactant). These original stock solutions were diluted by a factor of 8. Cells were grown in Dulbecco's Modified Eagle Medium (DMEM, Gibco) supplemented with 10% fetal bovine serum (FBS, Hyclone, UT) and 1% penicillin/streptomycin (P/S, Hyclone, UT) on a 96 well plate at a volume of 200 μL per well and a seeding density of 3500 cells/cm^2 in the presence of various concentrations of CaP coated particles, and a control with no particles. Nanoparticles were sonicated and vortex mixed before being added at volume of 20 μl per well. Cells were incubated at standard conditions (37°C, humidified, 5% CO_2,/95% air environment) for 1, 3 and 5 days before being read using the same procedure as that in the proliferation experiment above.

RESULTS AND DISCUSSION

TEM images (Figure 2) demonstrated that magnetite and maghemite nanoparticles with diameters ~ 20nm were successfully synthesized. Due to their magnetic properties and high surface energy, the nanoparticles formed aggregates.

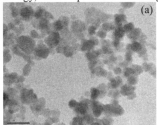 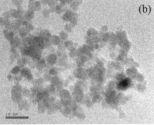

Figure 2: TEM images of (a) γ -Fe_2O_3 and (b) Fe_3O_4. The bar is 20nm.

Synthesized CaP had rod shapes similar to those in the previous research. Nano CaP was crystallized and had average sizes 50 nm long and 20 nm wide (Figure 3). TEM images also showed iron oxide nanoparticles embedded in the CaP particles.

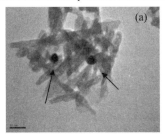 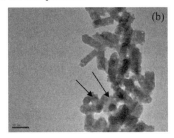

Figure 3: TEM images of (a) CaP coated γ -Fe_2O_3 and (b) Fe_3O_4. The bar is 50nm.

According to the VSM results (Figure 4), all nanoparticles showed properties of superparamagnetic materials. This meant that these nanoparticles do not retain any magnetism after removal of a magnetic field. As expected, the saturation magnetization of uncoated iron oxide nanoparticles (40 emu/g for Fe_3O_4 and 50 emu/g for Fe_2O_3) were much higher than that of CaP coated iron oxide nanoparticles. By increasing the ratio of iron oxide to CaP during the coating process, higher saturated magnetization of coated magnetic nanoparticles can be expected.

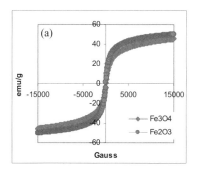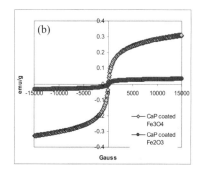

Figure 4: Magnetization curves of (a) uncoated and (b) CaP coated Fe_3O_4 and Fe_2O_3 nanoparticles as measured by VSM at room temperature. The results clearly showed superparamagnetic behavior of the nanoparticles.

Osteoblast proliferation tests conducted at 1, 3 and 5 days showed that Fe_3O_4 coated in the presence of CA and BSA increased OB density compare to the controls (no particles) (Figure 5). After 1 day, OB densities were similar in the presence of coated nanoparticles with CA and dextrant compared to the control. However, after 5 days, samples with CaP coated Fe_3O_4 added with CA and BSA showed significantly higher OB density compared to the control. Although further investigations should be made to understand the role of iron oxides and surfactants (CA and BSA) in the OB proliferation process, this study provided important evidence that CaP coated magnetic nanoparticles could improve bone cell function, and, thus, could be useful to treat osteoporosis and other bone diseases.

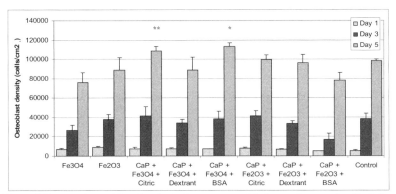

Figure 5: Increased osteoblast density in the presence of CaP coated magnetite (Fe_3O_4) and maghemite (Fe_2O_3) nanoparticles with surfactant CA, Dextrant and BSA after 1, 3 and 5 days of culture. Data = mean +/- SEM; N = 3. * p = 0.013 compared to control, ** p = 0.058 compared to the control (no particles).

The OB density experiments with various concentrations of CaP coated nanoparticles showed a very interesting dilution dependent trend. When the concentrations went from very dilute (#8 in the Figure 6) to very dense (#0 in Figure 6), the OB density decreased first, then increased and then decreased again. This trend was observed for nanoparticles with both surfactants CA and BSA. The trend indicated the existence of an optimum nanoparticle concentration. This information can be used to obtain most effective magnetic nanoparticle delivery system to treat bone diseases.

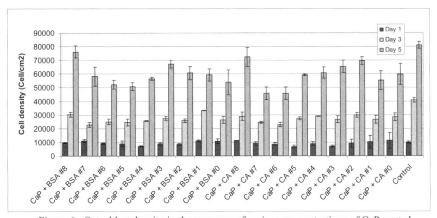

Figure 6: Osteoblast density in the presence of various concentrations of CaP coated magnetite (Fe_3O_4) nanoparticles with surfactants CA and BSA and control (without

nanoparticles) after 1, 3 and 5 days of culture. The number after # indicates the dilution factor from the original stock nanoparticle solutions. Data = mean +/- SEM; N = 3.

CONCLUSIONS

Magnetite and maghemite nanoparticles were successfully synthesized and characterized in this study using TEM. Further, these magnetic nanoparticles were coated with CaP in the presence of 3 different surfactants (CA, BSA and dextrant). After 5 days, the CaP coated magnetite in the presence of CA and BSA showed significantly higher osteoblast densities compared to controls (no particles). For future studies, cell experiments with CaP, iron oxide and surfactants separately should be performed to understand the contribution of each factor to the OB proliferation process. The study also demonstrated an important nanoparticle concentration dependence on OB density, which is useful for increasing the effectiveness of the drug delivery system.

.

ACKNOWLEDGMENTS

The authors thank Anthony McCormick for helping with TEM images, Jaemin Kim for helping with the VSM and the Hermann Foundation for funding.

REFERENCES

[1] A.K. Gupta, M. Gupta, "Synthesis and surface engineering of iron oxide nanoparticles for biomedical applications". *Biomaterials.* **26** 2005, pp. 3995–4021.

[2] R.A. Pareta, E. Taylor, T.J Webster, "Increased osteoblast density in the presence of novel calcium phosphate coated magnetic nanoparticles". *Nanotechnology* **19** 2008, pp. 265101.

[3] Y.S. Kang, S. Risbud, J.F. Rabolt, P. Stroeve, "Synthesis and characterization of nanometer-size Fe3O4 and γ -Fe2O3 particles". *Chem. Mater.* **8** 1996, pp. 2209-2211.

[4] W. Suchanek, M. Yashima, M. Kakihana, M. Yashimura. "Hydroxyapatite/hydroxyapatite-whisker composites without sintering additives: Mechanical properties and microstructural evolution". *J Am Ceram Soc* **80** 1997, pp. 2805–2813.

Mater. Res. Soc. Symp. Proc. Vol. 1140 © 2009 Materials Research Society   1140-HH05-34

Raman Spectroscopy of Defected Griseofulvin in Powders and Films

A. Żarów[a], W. Wagner[a], E. Lee[c], F. Adar[c], B. Zhou[b], R. Pinal[b] and Z. Iqbal[a*]

[a] Department of Chemistry, New Jersey Institute of Technology, Newark, New Jersey
[b] Department of Pharmacy, Purdue University, West Lafayette, Indiana
[c] Horiba Jobin Yvon Inc., 3880 Park Avenue, Edison, New Jersey
*Corresponding author: iqbal@njit.edu

ABSTRACT

Raman spectroscopy is used to investigate possible defect formation in the pharmaceutical material, Griseofulvin, by milling. The appearance of a broad background in the spectra recorded with 785 nm excitation suggests the formation of defects which fluoresce in the near-infrared spectral region. Three different methods were used to anneal the crystalline defects formed in the powders. Exposure to high level of humidity, acetone dissolution followed by recrystallization, and laser irradiation, were all found to be effective and showed partial reduction of the induced defects and reduction of the background intensity. Detailed examination of the low frequency Raman spectra associated with molecular deformation, rocking and wagging modes in milled and reference samples of Griseofulvin, also showed significant relative intensity differences in the Raman lines. Thin film strips prepared with Griseofulvin, sodium dodecyl sulfate, hydroxypropylmethyl cellulose, and glycerol were also examined by chemical micro-Raman imaging coupled with multivariate analysis to determine the uniformity of the drug distribution and the presence or absence of defects on the drug particles.

INTRODUCTION

Milling is widely used by the pharmaceutical industry as a way to reduce the particle size of active pharmaceutical ingredients (API's), excipients such as lactose, hydroxypropylmethyl cellulose (HPMC) or other components. Size reduction leads to better dissolution properties and increase in surface area of the particles. During the milling process, however, mechanical stress can induce changes in the crystal lattice affecting physical and chemical properties of the active ingredient. These changes can cause crystal defects and long-range disorder in atomic positions resulting in the formation of amorphous phases. Most of the first order lattice or inter-molecular modes occur at very low Raman frequencies below 200 cm^{-1} [1] whereas molecular deformation, rocking and wagging modes lie in the frequency range below 1000 cm^{-1} [2]. Any changes in peak positions and intensities or appearance of new lines could indicate the formation of defects, an amorphous phase or a new structure [3,4]. In addition, the presence of background intensity may be associated with disorder-induced scattering [5,6] in the Raman spectrum which would be independent of the excitation source used. However, if the background varies with excitation source, it is likely to be associated with fluorescence from induced-defects.

In the present study, crystalline defects were induced in a powder sample of Griseofulvin as observed by substantial changes in the intensity of the low frequency Raman lines and the appearance of a background with 785 nm near-infrared excitation. Raman imaging techniques have also been applied to film strips fabricated with Griseofulvin of different concentrations, in order to see if the film preparation techniques lead to defect formation. Since the crystalline state of an API affects its use in pharmaceutical applications as a drug processed into pills or capsules, it is necessary to develop effective, fast, and non-destructive characterization methods that would allow elimination of crystalline defects and potential polymorph formation [7-9].

EXPERIMENTAL DETAILS

Annealing study

All spectra for the annealing study were acquired using a Mesophotonics SE1000 Raman spectrometer calibrated to 2 cm^{-1} and a 785 nm excitation laser emitting with 300 mW power. 5 scans each 10 seconds long were taken to give a total of 50 second laser exposures for each spectrum acquired. The humidity study was performed in a sealed glass desiccator with the drying agent removed. The milled Griseofulvin sample along with 50 ml of hot de-ionized water and a humidity meter were placed in the desiccator. Humidity levels were maintained at 70-80% and the water was re-heated whenever the humidity level dropped below 70%. The ambient temperature in the desiccator was maintained between 21°-23° C. Small amounts of sample were removed periodically for measurement and were disposed after measurement rather than returned to the rest of the sample due to concerns that the laser exposure involved in the measurement would accelerate the annealing process. Samples were also tested at various stages of the drying process to determine what, if any, effect this variable might have on the results.

For the acetone dissolution study a small amount of the milled sample was completely dissolved in acetone. The solution was then heated to boiling on a hot plate until most of the acetone had boiled away, and then the damp solid which remained was allowed to dry in a hood to prevent degradation of the solid from heat on the hot plate. The sample was then analyzed after two hours of drying in air.

For the laser exposure study, a small portion of each sample was exposed to constant laser light for a total of 120 minutes, with spectra being obtained periodically. Due to the convenience of the experiment being performed in the spectrometer sampling chamber itself, the laser illumination was not disturbed at all during this time.

Confocal Raman Spectroscopy

To show differences between milled and reference Griseofulvin samples, individual spectra for each were collected using a Jobin Yvon/Horiba LabRAM micro-Raman spectrometer, with 632.5 nm excitation from a He-Ne laser and a 100x microscope objective with a spatial resolution of 1 μm. Each sample measurement corresponded to an average of 10 scans, each 10 seconds long, totaling 100 seconds in scanning time.

Confocal Raman Imaging of Pharmaceutical Films

Pharmaceutical thin film strips were investigated to determine the distribution of the embedded Griseofulvin drug, which had not previously been exposed to milling. Two pharmaceutical film strips were investigated in this study: film labelled 12C composed of 9.30% API (Griseofulvin), 69.74% hydroxypropyl methylcellulose (HPMC), 0.02% sodium dodecyl sulfate surfactant (SDS), and 20.94% glycerol; and film labelled 3-1 composed of 57% API and 47% HPMC. Chemical imaging was performed with a Jobin Yvon/Horiba LabRAM Aramis confocal Raman microscope using 632.8 nm He/Ne laser excitation and a 100x objective with a spatial resolution of 1 μm.

RESULTS AND DISCUSSION

The Raman spectral data in Figures 1(a) to Figure 1(c) taken using 785 nm excitation clearly show a strong and broad background in the milled Griseofulvin samples. On the assumption that this strong background is associated with the formation of defects on milling [3,4], the following studies were undertaken to investigate the annealing of these defects.

Humidity study:

The humidity study demonstrated at least partial annealing of crystalline defects in the milled sample and minimal change within the reference sample (see Figure 1a). The milled sample showed a reduction in background intensity in counts/second units (overall intensity minus relative intensity) at the location of the peak at 649 cm^{-1} from the original intensity value of 41,709 counts/sec to a value of 30,038 counts/sec, after two days of exposure to high humidity. However, the background intensity remained near 30,750 counts per second after seven days of exposure to high humidity, suggesting that almost all of the annealing occurred at least reasonably close to the initial exposure and that further exposure may not have any more significant effect. This demonstrated a 28.0% decrease in background intensity after two days of exposure, although it was not possible to quantitatively link the reduction in background intensity to a value for the percentage of defects annealed. Relative intensity (baseline-corrected intensity) remained almost constant through the study, demonstrating no loss of the Griseofulvin Raman signal in this process.

(a) (b) (c)

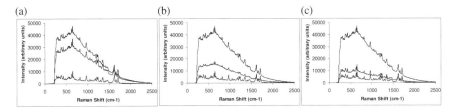

Figure 1. Spectral comparison for all annealing experiments: humidity (a), acetone (b), and laser exposure (c). In each figure top spectrum represents control milled Griseofulvin sample, middle spectrum – milled Griseofulvin sample after treatment, and bottom – control, Sigma-Aldrich Griseofulvin sample.

Acetone Dissolution Study:

The acetone dissolution study showed partial annealing of the defects in the milled sample as shown in Figure 1b. The milled sample showed a reduction in background intensity for the peak at 649 cm^{-1} from the original value of 41,709 to 15,393 counts/sec. This represents a 63.1% decrease in background intensity. One area of concern however with this process is a decrease in the relative intensity of the spectrum with a decrease in the relative intensity of the peak at 649 cm^{-1} from 5869 to 2782 counts/sec, which represents a 52.6% loss of the Griseofulvin Raman signal at that peak. This could be explained by the Griseofulvin solvate being still present due to incomplete drying.

Laser Exposure Study:

The laser study showed almost complete annealing of the amorphous defects in the milled sample after only two hours of exposure, see Figure 1c. Background intensity was reduced for the peak at 649 cm^{-1} from the original value of 41,709 to 8003 counts/sec after 120 minutes of laser exposure, a reduction of 80.8%. Relative intensity remained constant with only minor deviation throughout the process, demonstrating full maintenance of the Griseofulvin Raman signal and indicating that no sample degradation was occurring. The most dramatic annealing occurred very quickly with 70.0% of the observed annealing occurring within the first 10 minutes, and the annealing rate dropping throughout the process. This process shows great promise in potential pharmaceutical applications due to the quick rate of annealing, maintenance of the API's Raman signal, high annealing percentage, low cost of use, and ease of developing into an automated process.

Confocal Raman Spectroscopy and the Milled Griseofulvin Samples

Raman spectra of milled Griseofulvin and control Griseofulvin from Sigma-Aldrich obtained excited with 632.8 nm laser radiation, are shown in Figures 2(a) and 2(b). Significant differences can be observed when the two spectra are compared and most of these changes are

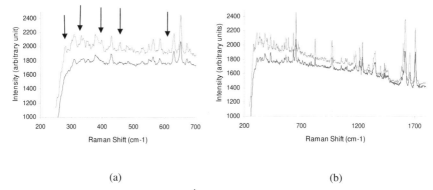

(a)   (b)

Figure 2: (a) Low frequency 250 to 700 cm^{-1} Raman spectra of Griseofulvin. Top - control sample of Griseofulvin from Sigma-Aldrich, bottom – milled sample of Griseofulvin. (b) Raman spectra in the 200 to 1800 cm^{-1} region of milled Griseofulvin (bottom) and control sample of Griseofulvin from Sigmal-Aldrich (top).

evident in the low frequency region from 250 to 550 cm^{-1}, where lines associated with molecular deformation/rocking/wagging modes and their combinations with the lattice modes of the Griseofulvin molecule occur. The main differences indicated in Figure 2(a) correspond to substantial loss in intensity of peaks at 277, 356, 382, 436, 454, 559, and 579 cm^{-1}. Some of these lines are indicated by arrows in the figure. An additional key observation is the sizable decrease of the broad background that was observed in the 785 nm excited Raman data discussed earlier. The fact that the background has different intensities at different excitation wavelengths suggests that it is associated with fluorescence, which is likely to arise from milling-induced defects.

Chemical Imaging of Griseofulvin-containing Film Strips

Distribution of Griseofulvin particles embedded in a pharmaceutical film strip is crucial to determining the optimal preparation conditions and monitoring any morphological changes during film strip fabrication. Figures 3(a) and 3(b) represent chemical images of film strips: 3-1 (57% API and 43%HPMC) and 12C ((9.3% API, 69.7% HPMC, 20.9% GLY and 0.02% SDS), respectively. Chemical images correspond mostly to the API particles, shown in white for film labelled 3-1 and film labelled 12C, and HPMC shown in black. The shades of white indicate API scattering intensity which changes due to topology of the film strip and uneven HPMC coverage. Each chemical map was further characterized by multivariate chemometric analysis to obtain improved API distribution (not shown here). The mapping data indicate no chemical changes of API and show drug agglomerations into larger particle regions from 5 to 10 μm. There are also smaller particles of size less than 5 μm.

(a)   (b)

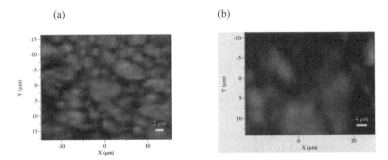

Figure 3: (a) XY chemical image of Griseofulvin distribution in film strip 3-1 (57% API and 43%HPMC); and (b) XY chemical imaging of Griseofulvin distribution in film strip 12C (9.3% API, 69.7% HPMC, 20.9% GLY and 0.02% SDS).

CONCLUSIONS

It is suggested that milling of Griseofulvin results in the formation of defects which fluoresce in the near-infrared region of the optical spectrum. More detailed spectroscopic investigations are planned to investigate the molecular structure of the defects formed. Out of the various methods used to anneal the defected structure formed, laser exposure proved to be the most effective by minimizing required duration and maximizing the desired result. The study also showed that Raman spectroscopy can be a powerful tool in investigating crystalline defects in pharmaceutical compounds. In addition to fluorescence, the defects caused substantial changes in the intensity of the lower frequency Raman lines of Griseofulvin, which are associated with molecular deformation, rocking and wagging modes, and their combinations with the lattice modes. The results of both the humidity and laser exposure studies have yielded results which can probably be developed into practical applications in pharmaceutical processing and quality control, and should be investigated further in future studies.

ACKNOWLEDGMENTS

A.Z., B.Z., R.P. and Z.I. would like to acknowledge support from the NSF-Engineering Research Center on Structured Organic Particulate Systems (EEC-0540855). W.W. was supported by an NSF-REU grant (EEC-0552587).

REFERENCES

1. B.A. Bolton and P. N. Prasad. "Laser Raman Investigation of Pharmaceutical Solids: Griseofulvin and Its Solvates". Jour. Pharm. Science **70**, 789 (1981).
2. R. L. McCreery. *Raman Spectroscopy for Chemical Analysis.* John Wiley & Sons, Inc.: New York, 2000.
3. T. Feng, R. Pinal, and M.T. Carvajal. "Process Induced Disorder in Crystalline Materials: "Differentiating Defective Crystals from the Amorphous Form of Griseofulvin". Jour. Pharm. Science, **97**, 3207 (2007).
4. S.P. Chamarthy and R. Pinal, "The Nature of Crystal Disorder in Milled Pharmaceutical Materials". Colloids and Surfaces A: Physicochem. Eng. Aspects **331**, 68 (2008).
5. M. A. Pimenta, G. Dresselhaus, M. S. Dresselhaus, L. G. Cançado, A. Jorio and R. Saito. "Studying Disorder in Graphite-Based Systems by Raman Spectroscopy". Phys. Chem. Chem. Phys. **9**, 1276 (2007).
6. Z. Iqbal. "Basic Concepts of Structural Phase Transitions", in *Magnetic Resonance of Phase Transitions* (Eds: F.J. Owens, C.P. Poole,Jr and H.A. Farach), Academic Press, New York (1979) p.1.
7. B. De Gioannis, P. Jestin, and P. Subra. "Morphology and Growth Control of Griseofulvin Recrystallized by Compressed Carbon Dioxide as Antisolvent". J. Cryst. Growth **262**, 519 (2004).
8. A. Docoslis, K. L. Huszarik, G. Z. Papageorgiou, D. Bikiaris, A. Stergiou and E. Georgarakis. "Characterization of the Distribution, Polymorphism, and Stability of Nimodipine in Its Solid Dispersions in Polyethylene Glycol by Micro-Raman Spectroscopy and Powder X-Ray Diffraction". AAPS Jour. **9**, E361 (2007).
9. S-L.Wang, S-Y. Lin, and Y-S. Wei. "Transformation of Metastable Forms of Acetaminophen Studied by Thermal Fourier Transform Infrared (FT-IR) Microspectroscopy". Chem. Pharm. Bull. **50**, 153 (2002).

Hydrogels and Material-Tissue Interactions

Mater. Res. Soc. Symp. Proc. Vol. 1140 © 2009 Materials Research Society   1140-HH07-07

Control of Adult Stem Cell Function in Bioengineered In Vitro Niches

Matthias P. Lutolf[1] and Helen M. Blau[2]
[1]Institute of Bioengineering, Ecole Polytechnique Fédérale de Lausanne, CH-1015 Lausanne, Switzerland
[2]Baxter Laboratory in Genetic Pharmacology, Stanford University School of Medicine, Stanford, CA 94305-5175, U.S.A.

ABSTRACT

Blood stem cells, also termed hematopoietic stem cells (HSC), are used clinically to treat numerous blood cancers. However, due to the low numbers of HSCs that can be isolated from the bone marrow of donors, broad application of this treatment is severely hindered. A robust method is needed that enables *in vitro* propagation and expansion of HSCs. In addition, as soon as HSCs are removed from their bone marrow niches and put in culture, they begin to differentiate and lose their multipotency. Therefore, a different experimental approach is required to control HSC function *in vitro*. We have developed a novel microfabrication process to generate synthetic hydrogel substrates comprised of arrays of microwells that can be locally functionalized with desired stem cell regulatory proteins in simplified 'artificial niches'. Time-lapse microscopy of single cells in conjunction with image analysis was employed to quantify changes in the proliferation kinetics of individual HSCs in response to specific niche proteins. In combination with subsequent *in vivo* transplantation assays, we identified proteins that maintain stem cell multipotency *in vitro* by controlling self-renewal divisions.

INTRODUCTION

A sparse population of hematopoietic stem cells is responsible for the production of blood for the entire life-span of an organism [1]. *In vivo*, the self-renewal and differentiation capacity of HSCs is maintained by a complex stem cell microenvironment, termed niche, that is comprised of diverse extracellular molecules [2]. In the absence of a niche, such as occurs during growth in conventional culture environments (i.e., plastic culture dishes), HSCs tend to rapidly specialize and lose their stem cell properties.

Although a variety of HSC niche components have been identified, the specific function of these signaling molecules remains poorly understood or is a matter of debate. In part, this can be attributed to the difficulties of studying the function of multifactorial stem cell niches *in vivo*. In addition, analyses of HSCs *in vitro* in bulk cultures have been hindered by the limited purity and thus unavoidable 'heterogeneity' of stem cell populations [3]. In the resulting mixed cultures, the behavior of enriched stem cells is often masked by the presence of highly proliferative progenitors, the specialized derivatives of stem cells.

We reasoned that microwell arrays [4-6] could be fruitfully applied to stem cell biology, providing a means to overcome HSC heterogeneity and shedlight on the effects of specific niche proteins on the function of single HSCs. Since in the niche HSCs are in contact not only with soluble proteins but also membrane components provided by support cells and insoluble extracellular proteins, we developed a novel micropatterning technique to mimic such interactions by presenting appropriately oriented tethered proteins to cells in arrays of hydrogel microwells. We show that stem cell self-renewal and long-term blood reconstitution can be induced by exposure of single HSCs to either soluble or tethered proteins presented in these highly simplified artificial stem cell niches.

EXPERIMENTAL

We selected poly(ethylene glycol) (PEG)-based hydrogels formed via conjugate addition reactions (Fig. 1A) as microwell substrates due to our previous successful use of this biomaterial in various cell culture and tissue engineering applications [7]. In contrast to tissue culture plastic or many previously developed microwell arrays, these PEG networks are soft (elastic modulus in the range of hundreds of Pa [8]) and have a relatively high water content (typically 95-98%). We therefore reasoned that PEG hydrogels could simulate the soft and hydrated microenvironment of HSCs in the bone marrow. In order to recapitulate HSC-niche interactions *in vitro* in a near-physiologic fashion, we developed a novel technique to spatially pattern and immobilize proteins onto our hydrogel matrices. Protein tethering was achieved by attaching a heterofunctional PEG linker to a protein of interest and then crosslinking this complex into the gel network (Fig. 1A). To ensure site-selectivity in protein immobilization, we focused on engineered Fc-chimeric proteins that could be linked via binding to an intermediate auxiliary protein, ProteinA, that contains four high-affinity binding sites ($K_a = 10^8$/mole) for the Fc-region of human, mouse and rabbit immunoglobulins.

Protein-functionalized hydrogel microwell arrays were fabricated by a multistep soft lithography process (Fig. 1B). Using standard photolithography, microwell patterns were first etched into silicon onto which liquid thermocurable poly(dimethylsiloxane) (PDMS) was polymerized (not shown here). To specifically functionalize gels and immobilize proteins only at the bottom of microwells rather than homogeneously distributing proteins across the entire array ('bulk modification'), PEG-functionalized ProteinA was adsorbed onto the posts of the PDMS template (Fig. 1B, step 1,2), a gel precursor solution was pipetted onto the ProteinA-adsorbed microstructured PDMS surface (step 3), and polymerization conducted between two hydrophobic glass slides separated by spacers (step 4). In this manner, the topographic pattern and the protein pattern were transformed onto the polymerizing gel surface. The resulting PEG hydrogel microwell arrays were peeled off, washed thoroughly and left to swell overnight in PBS. Before cell culture, Fc-chimeric proteins of choice were incubated and selectively bound to tethered ProteinA (step 5). Hydrogel microwell surfaces selectively modified with regulatory proteins were then used to trap single HSCs and study their *in vitro* behavior over time (step 6).

Confocal microscopy experiments demonstrated the spatial control of protein immobilization afforded by the novel hydrogel microcontact printing process (Fig. 1C): Immobilized, FITC-labelled BSA, a model protein, was shown to be anchored on the bottom of individual microwells (right panel), rather than on the entire surface of the microwell array (left panel). In addition, immunostaining demonstrated the selective immobilization of an Fc-ligand and Fc-chimeric N-Cadherin via binding to ProteinA (Fig. 1D).

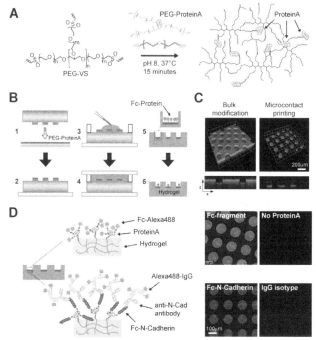

Figure 1. Fabrication of hydrogel niches for single hematopoietic stem cells. Covalent chemistry used to crosslink hydrogel networks functionalized with ProteinA (A). Novel process to locally functionalize hydrogel microwell arrays with proteins (B). Bulk versus local patterning of gel surfaces using FITC-BSA as a model protein (3D confocal micrographs of projection of 84 stacks acquired at a constant slice thickness of 1.8 µm) (C). Binding of Fc-ligand and Fc-chimeric N-Cadherin onto immobilized ProteinA revealed using immunostaining (negative controls: microwell arrays not tethered with Protein A or treated with isotype control primary antibody) (D). Reprinted with permission from [9]. Copyright 2009 Royal Society of Chemistry.

RESULTS AND DISCUSSION

The microfabricated platform was designed to be amenable to live-cell microscopic analyses of hundreds to thousands of single HSCs exposed to a desired protein, and compared simultaneously with HSCs exposed to other proteins. When seeded at low density onto microwell arrays, mouse hematopoietic stem cells that were isolated and purified using established fluorescence activated cell sorting (FACS) protocols (details described in [9]), sedimented by gravity as single cells to the microwell bottom, where they remained confined and started to divide (Fig. 2A). Even highly migratory HSCs were not able to escape from the microwells, facilitating long-term live-cell microscopic analyses (Fig. 2).

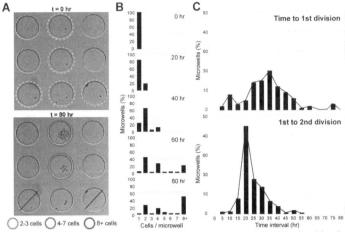

Figure 2. Time-lapse videomicroscopy to quantify single stem cell proliferation kinetics. Upon seeding onto hydrogel arrays, stem (or progenitor) cells sediment to the bottom of microwells (ca. 130μm diameter) where they are trapped and can be tracked over time (A). Image analysis reveals the distribution of cells per microwell over the course of the experiment (B), as well as the distribution of the times to the 1st division and times between 1st and 2nd division (C). Wells into which more than one cell sedimented t=0,were eliminated from analyses retrospectively (A,bottom). Adapted with permission from [9]. Copyright 2009 Royal Society of Chemistry.

Time-lapse microscopic analyses reveal single cell proliferation kinetics

We used automated time-lapse microscopy to assess the single cell proliferation kinetics of entire stem or progenitor cell populations. 200-300 single cells were seeded per well of a 96-well plate, each containing 400 microwells. Proliferation was scored as the distribution of cells per microwell at defined time intervals (Fig. 2B). We also measured the time between divisions for each individual cell of a given population (Fig. 2C). These experiments revealed that cells of different phenotypes, for example stem cells and progenitor cells, exhibited marked differences in division kinetics (not shown here). Progenitor cells were highly proliferative, while stem cells displayed slow division kinetics

Screening of niche components reveals distinct changes in proliferation

We systematically tested the effects of selected soluble and immobilized niche proteins on the proliferation kinetics of single stem cells. Strikingly, the addition of single proteins to the basal medium (serum-free culture medium supplemented with only two growth factors) markedly altered proliferation kinetics as shown in the histogram depicting the relative proportions of microwells with non-dividing cells (1 cell), slow dividing clones (2 cells), fast dividing clones (≥4 cells) or asynchronously dividing clones (3 cells) (Fig. 3). Four distinct proliferative patterns were observed. Representative proteins that induced the three patterns that differed from basal (Type I) were selected for further analysis (indicated by arrows), Wnt3a (Type II), TPO (Type III), and N-Cad (Type IV), of which the first two are soluble and the last is tethered.

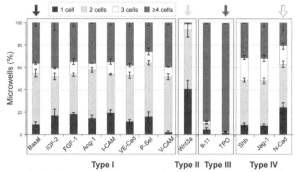

Figure 3. Screening for niche protein components that induced changes in single cell proliferation kinetics. Time-lapse microscopy of single HSCs exposed to individual (soluble and immobilized) protein components reveal significant changes in proliferation kinetics and synchrony. Adapted with permission from [9]. Copyright 2009 Royal Society of Chemistry.

Transplantation experiments reveal maintenance of stem cells after protein exposure

The disparate proliferation behaviors observed with TPO, Wnt3a and N-Cad suggested that single HSCs cultured in the presence of these three factors might have different biological properties. We tested whether the cells exposed to these proteins differed with respect to multipotency assessed by long-term blood reconstitution [3]. 100 HSCs were seeded in microwell arrays, exposed to TPO, Wnt3a, N-Cad or basal medium alone in culture for 4 days and all progeny transplanted into mice lethally irradiated to deplete their own stem cells (see [9]).

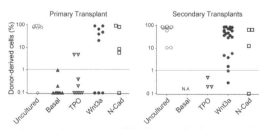

Figure 4. Transplantation experiments to test blood reconstitution potential of microwell-cultured HSCs. Cells exposed to basal conditions and TPO, a protein that increased proliferation kinetics, lost multipotency. By contrast, proteins that slowed down proliferation (Wnt3a) or led to asynchronous divisions (N-Cad) showed robust blood reconstitution. Adapted with permission from [9]. Copyright 2009 Royal Society of Chemistry.

After 6 months, a high efficiency of blood reconstitution (indicated here by the percentage of genetically marked donor-derived cells) was obtained in mice transplanted with

HSCs that were cultured in the presence of Wnt3a or N-Cad, whereas a low efficiency of reconstitution was obtained for the basal medium (control) and TPO (Fig. 4, left panel). These differences were even more pronounced upon secondary transplantation of HSCs from reconstituted mice into lethally irradiated recipients (Fig. 4, right panel). These results show that exposure to Wnt3a or N-Cad *in vitro* in hydrogel microwells leads to retention of stem cell function. In addition, both the rate and synchrony of stem cell division induced by single extrinsic factors *in vitro* correlated with *in vivo* HSC reconstitution potential in mice, suggesting that these characteristics could serve as predictors of maintenance of stem cell function.

CONCLUSIONS

A dissection of the regulatory role of specific signals within a complex stem cell niche is crucial for developing therapeutic strategies using adult stem cells. Our minimalist approach, a 'deconstruction of the niche', enabled the predominant role of specific proteins to be evaluated in the absence of the complexity of the *in vivo* microenvironment or *in vitro* co-cultures. Here we show that stem cell function can be maintained following cell division in culture. Further studies using this platform will discern the protein combinations necessary to expand the HSC population for clinical applications and will apply the platform to muscle cells [10].

ACKNOWLEDGMENTS

We are grateful to Dr. Regis Doyonnas, Karen Havenstrite and Kassie Koleckar who significantly contributed to this work. We thank our collaborators Drs. M. Textor and M. Dusseiller, ETH Zurich, for the generous gift of PDMS replicas, D. Klang, ETH Zurich for help with the development of microcontact printing. In the Blau laboratory, we thank P. Kraft for help with micromanipulation, Dr. P. Gilbert for help with microwell array fabrication. M.P.L. was supported by fellowships from the Swiss National Science Foundation and the Leukemia and Lymphoma Society. This work was supported by the Baxter Foundation, a Fullbright Senior Specialist Award and NIH grants AG009521, AG020961, AG024987 to H.M.B.

REFERENCES

[1] S. H. Orkin, L. I. Zon, *Cell* **132**, 631-644 (2008).
[2] D. T. Scadden, *Nature* **441**, 1075-1079 (2006).
[3] D. Bryder, D. J. Rossi, I. L. Weissman, *American Journal of Pathology* **169**, 338-346 (2006).
[4] D. Falconnet, G. Csucs, H. M. Grandin, M. Textor, *Biomaterials* **27**, 3044-3063 (2006).
[5] A. Khademhosseini, R. Langer, J. Borenstein, J. P. Vacanti, *PNAS* **103**, 2480-2487 (2006).
[6] G. H. Underhill, S. N. Bhatia, *Current Opinion in Chemical Biology* **11**, 1-10 (2007).
[7] M. P. Lutolf, J. A. Hubbell, *Nat Biotechnol* **23**, 47-55 (2005).
[8] M. P. Lutolf, G. P. Raeber, A. H. Zisch, N. Tirelli, J. A. Hubbell, *Advanced Materials* **15**, 888-893 (2003).
[9] M. P. Lutolf, R. Doyonnas, K. Havenstrite, K. Koleckar, H. M. Blau, *Integr. Biol.* DOI:10.1039/B815718A (2009).
[10] A. Sacco, R. Doyonnas, P. Kraft, S. Vitorovic, H.M. Blau, *Nature* **456**, 502-506 (2008).

Mater. Res. Soc. Symp. Proc. Vol. 1140 © 2009 Materials Research Society   1140-HH06-19-DD03-19

Hyaluronic acid –gelatin fibrous scaffold produced by electrospinning of their aqueous solution for tissue engineering applications

Ying Liu[1], Richard A.F.Clark[2], Lei Huang[3], and Miriam H.Rafailovich[1]
[1]Department of Material Science and Engineering, Stony Brook University, Stony Brook, NY 11794-2275, USA
[2]Department of Biomedical Engineering, Stony Brook University, Stony Brook, NY 11794-8181, USA
[3]Department of Condensed Matter Physics and Materials Science, Brookhaven National Laboratory, Upton, NY, 11793, USA

ABSTRACT

A three-dimensional (3D) hyaluronic acid (HA) and gelatin (Gtn) fibrous scaffold was successfully fabricated to mimic the architecture of natural extracellular matrix (ECM) based on electrospinning. Thiolated HA and Gtn derivatives, HA-DTPH and Gtn-DTPH, were synthesized and electrospun to form composite 3D fibrous scaffolds. In order to facilitate the fiber formation during electrospinning, Poly (ethylene oxide) (PEO) was added into the aqueous solution of HA-DTPH and Gtn-DTPH at an optimal weight ratio of 1:1.2. The electrospun HA-Gtn/PEO blend scaffold was subsequently cross-linked through Poly (ethylene glycol)-diacrylate (PEGDA) mediated conjugate addition. PEO was then extracted in deionized (DI) water to obtain an electrospun HA-Gtn fibrous scaffold. Adult human dermal fibroblasts (AHDFs) were seeded on HA-Gtn fibrous scaffold for 24h in vitro. Laser scanning confocal microscopy (LSCM) revealed that the AHDFs attached to the scaffold and spread, which suggests potential applications of HA-Gtn firbous scaffolds in tissue regeneration.

INTRODUCTION

Tissue engineering requires the utilization of a porous biodegradable scaffold to replicate the natural ECM, which serve to organize cell in space, to provide them with environmental signals and to direct site-specific cellular regulation. Hyaluronic acid (HA) is a natural occurring linear polysaccharide and a major constituent of the extra cellular matrix (ECM). It plays a key role in stabilizing and organizing ECM, regulating cell adhesion and mobility, and mediating cell proliferation and differentiation[1]. Even more, HA-based hydrogels are ideal materials for soft-tissue engineering because of their unique rheological properties and complete biocompatibility[2].

Previous work has described the fabrication fibrous scaffold of a low molecular weight thiolated HA derivative, 3, 3'-dithiobis-(propanoic dihydrazide)-modified HA (HA-DTPH). The NIH 3T3 fibroblasts were seeded on the FN-adsrobed HA-DTPH to study the cell morphology [3, 4]. However, we found that human dermal fibroblasts, which are the primary cells involved in cutaneous reparative process, were not able to attach to such scaffold, even with fibronectin absorption.

In this study, a thiolated HA and gelatin (Gtn) derivative, HA-DTPH and Gtn-GTPH, was synthesized to fabricate fibrous scaffolds. Poly (ethylene oxide) (PEO) was blend with HA-DTPH and Gtn-DTPH as a viscosity modifier to facilitate the fiber formation during electrospinning. A uniform HA-Gtn/PEO fibrous scaffold without beads was fabricated, which was then cross-linked through poly (ethylene glycol)-diacrylate (PEGDA)-mediated conjugate addition. Adult human dermal fibroblasts (AHDFs) were successfully grown on HA-Gtn fibrous scaffold and dendric morphology of the cells was observed.

EXPERIMENT

Electrospinning solution preparation

The synthesis and characterization of thiolated HA and Gtn derivatives, HA-DTPH and Gtn-DTPH was described in an early article[5]. The free thiols content in HA-DTPH and Gtn-DTPH was found to be approximately 42% and 37%.The molecular weight of HA-DTPH has been determined to be M_w =158kDa and M_n =78kDa (M_w / M_n =2.03) and the molecular weight of commercially obtained Gtn-DTPH was reported to be~50kDa.

PEO powder was dissolved in Distilled phosphate-buffered saline (1 × dPBS, PH 7.4) at concentration of 2.5% or 3.0 %(w/v). HA-DTPH and Gtn-DTPH were then added into the PEO/dPBS solutions at a total concentration of 2.5% (w/v) (with weight ratio of 1:4) and dissolved using a vortex mixer (Vortex-genie 2, Scientific Industries, Inc.) until the solution became clear. Poly (ethylene glycol) diacrylate (PEGDA, M_w = 3400Da), as cross-linker, was dissolved in dPBS solution at concentration of 9.0%.

Fabrication of HA-DTPH and Gtn-DTPH composite fibrous scaffold

The procedure for fabricating HA-Gtn fibrous scaffolds is shown in Figure 1. For the fabrication of electrospun HA-DTPH and Gtn-DTPH composite fibrous scaffolds, the HA-DTPH, Gtn-DTPH and PEO blend solutions were loaded in a 5 ml glass syringe (Popper and Sons Inc., New Hyde Park, NY) attached to a syringe pump (KDS200, KD Scientific Inc., New Hope, PA), which provided a steady solution flow rate of 30 μ l/min during electrospinning. A high-voltage power supply (Gamma High Voltage Research, Ormond Beach, FL) was employed to apply a high potential of 15 kV between the needle and the a metal target, which was horizontally placed 10cm away from the tip of the needle. The electrospun fibers were deposited on aluminum foil or silicon wafer for form three-dimensional (3D) scaffold. PEGDA was subsequently added into the as-spun scaffold using an eppendorf micropipette. The blend scaffold was dried and cross-linked in air at room temperature. After 24h, the cross-linked scaffold was soaked into DI water for another 24h to extract PEO and form HA-Gtn hydrogels.

Surface morphologies of the electrospun scaffolds were characterized using SEM (LEO1550, LEO, Germany) at 10kV acceleration voltage and 8mm working distance. Swollen HA-Gtn scaffold samples were obtained by immersing the HA-Gtn in deionized (DI) water for 24h to reach equilibrium, freezing quickly in a freeze dryer (Consol 1.5, Virtis Inc., NY) at -40℃, followed by lyophilization. Samples were sputter coated with gold for 15s twice and loaded on aluminum stubs for SEM imaging. The fiber diameter distributions of HA-Gtn composite scaffolds were calculated by analyzing the SEM image using Image Tool (The University of Texas Health Science Center in San Antonio). A TEM (JEOL 3000F, Japan) was employed to investigate the interior structure of the electrospun fibers. TEM samples were obtained by directly electrospun the fibers onto a copper sample grid.

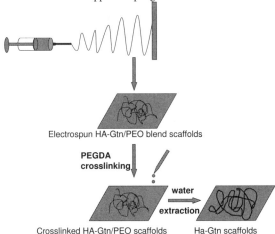

Electrospun HA-Gtn/PEO blend scaffolds

PEGDA crosslinking

water extraction

Crosslinked HA-Gtn/PEO scaffolds   Ha-Gtn scaffolds

Figure 1. Schematic illustration of the procedure of fabricating HA-Gtn fibrous scaffolds.

Cell morphology study

Primary AHDFs were obtained from Clonetics (San Diego, CA) and used between passages 6 and 13. The cells were routinely cultured in Dulbecco's modified eagle's medium (Sigma Chemical Co., St. Louis, MO, DMEM) supplemented with 10% fetal bovine serum (Hyclone, Logan, UT) and antibiotic mix of penicillin, treptomycin, and L-glutamine (full-DMEM), in a 37 ℃, 5% CO_2/ 95% air incubator.

The cross-linked HA-Gtn scaffolds were UV sterilized for 15min. The AHDF cells in full-DMEM were then seeded on electrospun HA-Gtn scaffold at a density of 1.25×10^4 cells/cm² and incubated at 37℃ with 5% CO_2 for 24h. As positive controls, cells were seeded at the same density on glass cover slips. After incubation for 24h, the supernatant was aspirated and the cultured cell sample was fixed with 3.7% (v/v) formaldehyde for 15min. The cells were then permeabilized with 0.4% Triton and stained with alexa-flour 488 phallodium (Invitrogen, Carlsbad, California) and Propidium Iodide (HPLC, Sigma Chemical Co., St. Louis, MO) for actin cytoskeleton and nucleus, respectively. The morphology of the cells on samples was visualized with a Leica TCS SPS laser scanning confocal microscopy (Leica micro-system Inc., Bannockburn, IL, LSCM).

DISCUSSION

Optimization of HA-Gtn/PEO weight ratio

To optimize the HA-Gtn/PEO weight ration, the ratio of HA-DTPH/Gtn-DTPH was fixed at 1:4 first, but ratio between the HA-Gtn and the PEO was changed from 1:1 to 1:1.2. At an HA-

Gtn/PEO weight ratio of 1:1 (Figure 2a), the elecrospun fibers showed a beads-on-string morphology. As the HA-Gtn/PEO weight ratio increased slightly from 1:1 to 1:1.2 (Figure 2b,c), beads disappeared and a uniform fibrous scaffold was obtained. We then selected the weight ratio of 1:1.2 as the optimal weight ratio to fabricate HA-Gtn fibrous scaffolds.

TEM picture in Fig.2d showed that no fine structure was found on the electrospun HA-Gtn/PEO blend fiber, which may mainly due to the homogeneous distribution of the HA-DTPH and Gtn-DTPH within the fibers. The contrast on the fiber was coming from the salt in the dPBS solutions, which was used as a solvent in the electrospinning process.

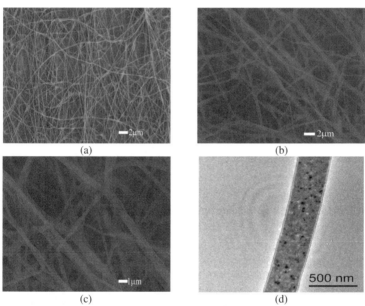

(a) (b)

(c) (d)

Figure 2. Morphologies of the electrospun HA-Gtn/PEO blend scaffolds with different weight ratios: (a) HA:Gtn=1:4, HA+Gtn=2.5%(w/v), PEO=2.5%(w/v); (b) HA:Gtn=1:4, HA+Gtn=2.5%(w/v), PEO=3.0%(w/v); (c) is the higher magnification of (b); (d) TEM image of a electrospun HA-Gtn/PEO blend fiber: HA:Gtn=1:4, HA+Gtn=2.5% (w/v), PEO=3.0% (w/v).

Fabrication of HA-Gtn fibrous scaffolds

In order to stabilize the HA-Gtn fibrous scaffolds in water, PEGDA was selected to cross-link the system, because of the rapid cross-linking rate and the controllable cross-linking density. Since PEO is chemically inert to PEGDA and can be dissolved in water, the as spun HA-Gtn/PEO blend scaffold was subsequently soaked in DI water to remove PEO and obtain an HA-Gtn fibrous scaffold. The morphologies change of HA-Gtn fibrous scaffolds after PEO extraction was shown in figure 3. Surface SEM images showed that the HA-Gtn scaffold still maintained a 3D fibrous structure after PEO extraction. Meanwhile, we found that some fibers fused together after PEO extraction and the shape of fibers was not as uniform as before PEO extraction. Using

Image Tool, we quantitatively analyzed the change of fiber diameter distributions before and after PEO extraction: before PEO extraction, more than 90% of fibers were within the diameter range between 200 to 600nm; After PEO extraction, the distribution of fiber diameters became much wider and 90% of fibers were within the diameter range between 500 and 1000nm. Some fibers even had larger diameter up to 1100nm.

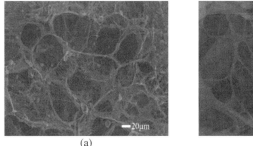 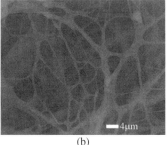
(a)   (b)

Figure 3. Morphological change of the HA-Gtn fibrous scaffolds after PEO extraction

Cell morphology on HA-Gtn fibrous scaffolds

LSCM was utilized to investigate the cell morphology of AHDF on the cross-linked HA-Gtn fibrous scaffold. As a control, AHDF cells were seeded on glass cover slips with the same cell density and cultured at the same condition. The typical flattened morphologies of AHDFs were observed on glass cover slips (figure 4a). While the on HA-Gtn fibrous scaffold, cells had been penetrate to different layer of the scaffold. Figure 4b shows reprehensive cells on the scaffold. They demonstrated dendric shape on the HA-Gtn fibrous scaffolds, and more filopodia extending out from the cell body (figure 4b).

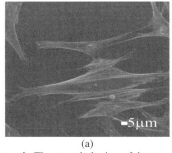 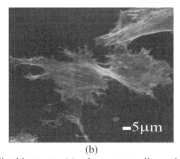
(a)   (b)

Figure 4. The morphologies of human dermal fibroblasts on (a) glass cover slip and (b) crosslinked HA-Gtn fibrous scaffold.

In this study, Thiolated HA and Gtn derivatives (HA-DTPH and Gtn-DTPH) were electrospun and cross-linked in air to give disulfide-cross-linked 3D porous scaffold. PEO was added into the HA-Gtn aqueous solution to facilitate the fiber formation during electrospinning. The as-spun HA-Gtn blend scaffolds were crosslinked by PEGDA. PEO was subsequently extracted in DI water to ultimately yield an HA-Gtn fibrous scaffold with range from 400 to 1100nm. AHDF cells migrated into the scaffold through the porous structure to form a 3D dendritic morphology inside the scaffold. These results suggest potential application of 3D HA-Gtn nanofibrous scaffolds in cell tissue regeneration.

ACKNOWLEDGMENTS

Financial support provided by M.H.R. grant. The author also would like to thank Prof. Glenn D. Prestwich and Dr. Anna Scott from the University of Utah to provide the HA-DTPH and Gtn-DTPH.

REFERENCES

1. Allison DD, Grande-Allen KJ, Tissue Engineering.**12**, 2131 (2006).
2. Shu XZ, Liu YC, Palumbo FS, Lu Y, Prestwich GD, Biomaterials. **25**, 1339 (2004)
3. Ji Y, Ghosh K, Shu XZ, Li BQ, Sokolov JC, Prestwich GD, Clark RA, Rafailovich MH, Biomaterials. **27**, 3782 (2006).
4. Ji Y, Ghosh K, Li BQ, Sokolov JC, Clark RAF, Rafailovich MH, Macromolecular Bioscience. **6**, 811(2006).
5. Shu XZ, Liu YC, Palumbo F, Prestwich GD, Biomaterials. **24**, 3825 (2003).

Mater. Res. Soc. Symp. Proc. Vol. 1140 © 2009 Materials Research Society   1140-HH06-17-DD03-17

Modified Alginate for Biomedical Applications

Soumitra Choudhary, Jason Reck and Surita R. Bhatia
Department of Chemical Engineering, University of Massachusetts, Amherst, MA 01003, USA

ABSTRACT

Hydrophobically modified alginate (HMA) was synthesized by attaching n-octylamine groups on to the alginate backbone at low pH. At or above a critical concentration, HMA forms a physical gel in aqueous media due to hydrophobic interactions. Unreacted guluronic units of alginate were further crosslinked with divalent cations, such as Ca^{2+}. Uniform gels were obtained by slow release of calcium ions from a calcium-ethylene diamine tetra acetic acid (Ca-EDTA) complex with the addition of D-glucono-δ-lactone (GDL). Mechanically stiffer gels of storage moduli ~ 100 kPa were obtained at a relatively low polymer concentration of 2 wt% in the crosslinked gels. Solubility of a model lipophilic drug, sulindac, in HMA was found to be greatly improved compared to neat alginate, presumably due to preferential uptake of the drugs by micelles formed by the hydrophobic moiety. Extended release of sulindac was observed for upto 5-6 days, probably due to stronger crosslinked alginate units surrounding the hydrophobic rich domains. Small angle x-ray scattering data suggest a strong influence of hydrophobic group on the nanometer-scale structure, especially in the uncrosslinked state.

INTRODUCTION

Hydrophobically modified polyelectrolytes are an important class of amphiphilic polymers widely used as thickeners, binders or gelling agents in the pharmaceuticals and food industries [1,2]. Hydrophobic groups attached to the polyelectrolyte backbone enable them to exhibit distinct characteristics, which are influenced by the interplay between hydrophobic association and electrostatic interaction. In suitable solvents, at and above critical concentration (semi-dilute regime) these amphiphilic polymers form intermolecular association strong enough to form reversible three-dimensional networks or gels. In the current study, we investigated hydrophobically modified alginate as a potential candidate for extended release of model lipophilic drugs and other biomedical applications.

Alginate is a linear naturally occurring copolymer of 1,4-linked β-D-manuronic (M) and α-L-guluronic (G) acid residue, usually derived from brown algae. Guluronic units of adjacent alginate chains can be crosslinked by diavalent cations such as Ca^{2+}, Ba^{2+}, etc. Since only G units can form a bond with the cations, the ratio of M/G is a critical parameter in determining the gel properties [3]. They are very popular in cell encapsulation because of mild gelling conditions and the ease with which many biologically relevant entities can be immobilized [4-6]. Hydrophobic modification of alginate was found to enhance the stability of gels in water and other biological buffers containing calcium chelators, due to the attractive association among hydrophobic moiety in aqueous media [7]. In this report, we have demonstrated that hydrogels made from HMA can be used to control the release of lipophilic drugs e.g. sulindac, in a simulated physiological condition. HMA was synthesized by forming amide linkages between n-octylamine and carboxylate group present on the alginate backbone. Ability to alter the degree of substitution of hydrophobic group gave us the extra handle to tune the rheological and mechanical properties of the system.

EXPERIMENT

Synthesis and characterization of hydrophobically modified alginate was published in details by Galant *et al.* [1]. In brief, 3 wt% aqueous alginate solution was acidified to pH ~ 3.4 by adding 0.1M HCl solution. To this solution the coupling agent 1-ethyl-3-(3-dimethylamino-propyl)carbodiimide was added and stirred for 5 minutes. Octylamine (n-C_8) was added in stoichiometric amounts and stirred for another 24 hrs at ambient condition. After the completion of reaction the polymer was precipitated in acetone, and separated by centrifuging. HMA was purified by washing again with acetone followed by dialysis against nanopure water for 7 days. Finally, the product was recovered by freeze drying.

Pure alginate or its modified version was gelled by *in-situ* release of calcium ions from Ca-EDTA by adding GDL [8]. This resulted in a more uniform gel than simply adding the ions from $CaCl_2$ solution. In a typical method of sample preparation, the dilute polymer solution was mixed with 0.3M Ca-EDTA for about 12 hrs. GDL was added to the mixture, which slowly hydrolyzes the Ca-EDTA and releases the ions. For rheological measurement the final mixture was quickly transferred to the couette geometry, and left for 48 hrs to complete the gelation before starting the frequency sweep in linear viscoelastic region. In this report, "sol" represents uncrosslinked sample irrespective of concentration, and was prepared by simply dissolving the required amount of polymer in water.

For drug release tests, 0.5 wt% of drugs were loaded into the 2 wt% HMA solution, stirred thoroughly for 24 hrs, then gelled in the same manner as described earlier. We also performed release studies on systems with pure alginate, and no-polymer; in these cases, the drug loading is kept at maximum possible concentration of 0.147 wt%. The samples were then sealed inside cellulose membranes (1000 MWCO) and transferred to continuously stirred phosphate buffer saline (PBS) maintained at 37^0C. The drug concentration in PBS buffer was monitored with UV-spectrometer at wavelength corresponding to peak intensity.

RESULTS

Figure 1 shows the small amplitude oscillatory shear data of HMA at molar degree of substitution of 30 (HMA-ds30). It can be seen from sol data (fig. 1-A) that even at low

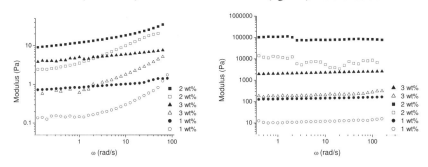

Figure 1: Frequency sweep of HMA-ds30 in water a) Sol, b) Gel. Closed symbols represent the elastic modulus, and open symbols the viscous modulus.

polymer content the profile is almost frequency independent, indicating that the sol behaves as an elastic solid. Since the pure alginate solution does not gel unless at higher polymer content [7], the gel like behavior exhibited by sol has to be due to intra-molecular association of hydrophobic group present in the alginate backbone. The elastic moduli ($G' \leq 10$ Pa) of the sol were found to be much lower than pure alginate (~ 1 kPa). This is probably due to polymer degradation during HMA synthesis, as also been confirmed by significant reduction in viscosity (data not shown) of HMA sol compared to alginate at similar concentration. The gels (fig. 1-B) at the same concentration however, showed significant improvement in the storage moduli (~ 100kPa). This can be attributed to synergistic effect of two different gelation phenomenons working in tandem to yield better properties. It is to be noted that the best property was found not at highest concentration of 3 wt% reported in this paper but at 2 wt%, indicating that there is an optimum concentration, beyond which the systems tend to phase separate. This behavior is observed for both sol and gel samples.

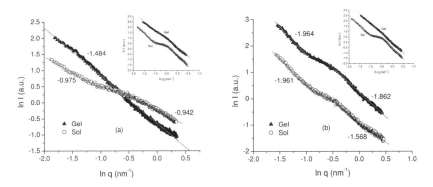

Figure 2: Aqueous SAXS profile of a) 1 wt% HMA-ds30, b) 2 wt% HMA-ds30. Some data are shifted vertically for clarity. (Inset) HMA-ds15 at respective concentration.

In order to determine structure-property relationship we used small angle x-ray scattering (SAXS) technique to explore the molecular architecture of the gel. Monochromatic x-ray beam at 0.1542 nm wavelength is passed through the samples to probe the network structure in the length scale of 4-80 nm. Using the scaling argument of the exponent from the value of fractal dimension (D) obtained from scattering ($I(q) \approx q^{-D}$), we were able to understand the structure of the final network [9,10]. From figure 2, we can conclude that the network is more compact in the case of gel compared to sol as can be inferred from higher values of exponent. The denser network explains the improved mechanical strength and better release profile, as we will see later. It can be seen from the figure that there are two distinct regions in SAXS profile of sol, which indicate ordering at two different length scale. The discontinuity may be due to hydrophobic modification as intermediate region is absent in pure alginate solution (data not shown). Similar behavior is observed in 2 wt% HMA-ds30 gel, however 1wt% gel (fig 2-A) and all gels at lower degree of substitution (HMA-ds15, see inset fig 2) exhibited a single slope indicating that the structure is preserved over the entire q range.

Drug release profiles of the anti-inflammatory drug sulindac with and without HMA are shown in figure 3. The drug release has been significantly improved with HMA system for both sol and gel state over sample with no-polymer and for neat alginate system. Although pure alginate seems to exhibit slower release, it must be emphasized that the drug loading is about 3.5 times less than the HMAs. The net driving force is expected to be lower in case of alginate and hence the positive effect of hydrophobic modification is two fold. However we did observe better profile with alginate compared to the no-polymer system, indicating that crosslinking has some leverage in drug release mechanism. More importantly we did not see any burst release for HMA, as in the case of the no-polymer solution, but a prolonged and controlled release over a period of about 5 days. As expected, release is better for HMA-gel because of higher mechanical strength and integrity, which enables them to retain the drug for much longer.

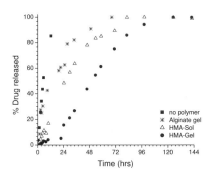

Figure 3: Drug release profile of sulindac in PBS at 37^{0}C.

Differences in the release rate for sol and gel can be attributed to a drug-hydrophobic "micelle" being surrounded by crosslinked unmodified-G units of HMA chains in case of gels, as shown schematically in figure 4. Whereas for sol, it is more appropriate to believe that the drug-polymer complex are been held only by reversible physical bond, which can swell much more quickly releasing the encapsulated drug out of the matrix in a shorter period of time.

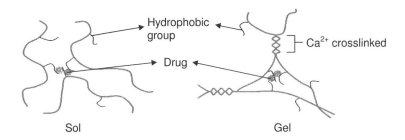

Figure 4: Schematic of drug-HMA distribution in sol and gel state.

CONCLUSIONS

We were able to successfully synthesize hydrophobically modified alginate for potential drug delivery applications. Hydrophobic modification of alginate has a significant effect on stability and mechanical properties of the gels. Physically, HMA gels were found to be very strong even at relatively low polymer content. The stronger gel directly translates to better release profile for a model hydrophobic drug. This study clearly demonstrates that the hydrophobic drug can be successfully stabilized in an aqueous medium, and its release can be controlled for an extended period of time. Since the developed biomaterial predominantly consists of water, there are immense possibilities that its applications can be extended to tissue engineering and cell encapsulation.

ACKNOWLEDGMENTS

Support for this work was provided by NSF-Institute of Cellular Engineering-IGERT program (DGE-0654128), ICE-REU program (EEC-0649041), NSF-MRSEC UMass Amherst (DMR-0213695), NSF-Career Grant (CBET-0238873), and NSF-Center for Hierarchical Manufacturing (CMMI-0531171). Authors would also like to thank Elam Prakash and Prof. S. Thayumanavan, Department of Chemistry, UMass Amherst for helping us in HMA synthesis.

REFERENCES

1. C. Galant, A. L. Kjoniksen, G. T. M. Nguyen, K. D. Knudsen, and B. Nystrom, *J. Phys. Chem. B*, **110**, 190 (2006).
2. K. D. Knudsen, R. A. Lauten, A. L Kjoniksen, and B. Nystrom, *Eur. Polym. J.*, **40**, 721 (2004).
3. A. D. Augst, H. J. Kong, and D. J. Mooney, *Macromolecular Bioscience* **6**, 623 (2006).
4. S. R. Bhatia, S. F. Khattak, and S. C. Roberts, *Current Opinion in Colloid and Interface Science*, **10**, 45 (2005).
5. K. Chin, S. F. Khattak, S. R. Bhatia, and S. C. Roberts, *Biotech. Progress*, **24**, 358 (2008).
6. M. K. E. McEntee, S. K. Bhatia, L. Tao, S. C. Roberts, and S. R. Bhatia, *J. Appl. Polym. Sci.*, **107**, 2956 (2008).
7. M. R. De Boisseson, M. Leonard, P. Hubert, P. Marchal, A. Stequert, C. Castel, E. Favre, and E. Dellacherie, *J. Colloid Interface Sci.*, **273**, 131 (2004).
8. X. Liu, L. Qian, T. Shu and Z. Yong, *Polymer* **44**, 407 (2003).
9. J. L. Higgins and H. C. Benoit, "Polymers and Neutron Scattering", Clarendon Press, Oxford (1994), pp. 168-174.
10. G. Beucage, *J. Appl. Cryst.*, **29**, 134 (1996).

Mater. Res. Soc. Symp. Proc. Vol. 1140 © 2009 Materials Research Society   1140-HH10-04

Tissue Integration of Two Different Shape-Memory Polymers with Poly(ε-Caprolactone) Switching Segment in Rats

B. Hiebl[1], D. Rickert[2], R. Fuhrmann[3], F. Jung[1], A. Lendlein[1], R.-P. Franke[1,3]

1: Centre for Biomaterial Development and Berlin Brandenburg Center for Regenerative Therapies (BCRT), Institute for Polymer Research, GKSS Research Centre GmbH, Teltow, Germany
2: University of Ulm, Department of Otolaryngology and Head and Neck Surgery, Ulm, Germany
3: University of Ulm, Central Institute for Biomedical Technique, Department of Biomaterials, Ulm, Germany

INTRODUCTION

One of the major challenges in biomaterial science is to achieve a strong connection between polymer-based biomaterials and the surrounding tissue guaranteeing a long-term-durable integration which resists mechanically induced stress like tensile loads and shear stress [1, 15, 17]. This biomaterial-tissue integration can be affected by the biofunctionality of the biomaterial in use [13, 14, 16]. For degradable polymers like poly[(L-lactide)-co-glycolide] it has been reported that the integration is influenced by the pH at the polymer-tissue interface [2, 4, 6, 11]. Also the formation of a thick fibrous capsule around the polymer implant can undermine polymer-tissue integration by disrupting the blood supply to the site of polymer-tissue interface. Insufficient blood supply to the site of interface disturbs necessary pH compensation and also the removal of degradation products. In addition, the surrounding tissues can not be supplied with nutrients without sufficient blood supply, and cells like monocytes/macrophages known to be necessary for tissue re-organisation processes.

Thus in the present study, we examined the integration of two different shape-memory polymers containing poly(ε-caprolactone) segments into the subcutis of a rat focussing on the pH change directly at the polymer-tissue interface and also on the characterisation of the periimplantary tissue nearby the polymer-tissue interface.

MATERIALS AND METHODS

Implants

The study was performed with two shape-memory polymers. A cross-linked AB-polymer network prepared from poly(ε-caprolactone) dimethacrylate (40 wt-%) and n-butylacrylate (60wt-%) (PCL/BA) [5] was compared with a physically cross-linked linear multiblock copolymer (PDC) consisting of poly(p-dioxanone) as hard segment (40 wt-%) and poly(ε-caprolactone) as switching segment [5]. Polymer samples used for the implants were in a discs format (a diameter of 12 mm and a thickness of 0.57 ± 0.02 mm).

Surgical procedure

All animal experiments for this study were approved by the regional board of Giessen and performed under specific pathogen-free conditions in the animal facility of the Phillips-University Marburg (Marburg, Germany). Adult Sprague-Dawley rats (310-360 g, Charles River, Germany) were maintained in isolated ventilated cages (one animal per cage, 12/12 light/dark-cycle). For implantation (PDC: group A; PCL/BA: group B; 15 animals/group) rats were premedicated with atropine (0.05 mg/kg) and anesthetized with a mixture of ketamine (75 mg/kg) and xylazine (12 mg/kg). An incision of 1.5 cm was made dorsomedial directly in front of the regio interscapularis and the implant was housed in the connective tissue of the subcutis. Additional 15 animals were sham operated (group C). After surgery, wound closure was performed by suturing (3-O Vicryl^{TM}, Ethicon). Bleeding was minimized by using a cauterizer in the surgical procedure. Every week after implantation, up to 5 weeks, 3 rats from each group were sacrificed by the use of CO_2. The skin of the implantation site was opened in the direction of the former incision, blood was swabbed, and the pH of the polymer-tissue interface was measured using an invasive pH electrode (MI-414 and MI-413, INC, Bedford, USA) and also indicator sticks (Merck, Neutralit$^{®}$ pH 5-10). There was no significant difference between the pH values measured by the electrode and the pH indicator sticks. After measuring the pH, the implants were explanted together with the surrounding tissue for further histological examinations.

Histology
Immediately after the explantation, each polymer-tissue sample was cryoconserved following the method described by Romeis [12]. Samples were put in 4 ml of embedding medium (Tissue Teck$^{®}$) on a cork plate. Thereafter the cork plate with the sample on top was placed on liquid nitrogen until the whole sample was frozen. The frozen samples were stored at -80 °C. For immunohistological examinations, cryosections (thickness: 5 μm; Leica CM 3050S) were prepared and they were subjected to hematoxylin-eosin (HE) staining and Carstairs staining according to the protocol of Romeis [12]. Five slides per sample were prepared and each slide was evaluated at five different fields of view.

Statistics
For all quantification analysis, the results were expressed as means ± standard deviation of at least three independent experiments. The significance of differences was assessed using two-tailed Student´s t-test for unpaired samples. Significance was assumed if p value was less than 0.05.

RESULTS
Already 3 weeks after implantation a strong connection between the PDC implant and the periimplantary tissue could be noted (fig. 1). In fact, this PDC-tissue connection and integration respectively was strong enough that the animal could be lifted only by holding the PDC disc up.

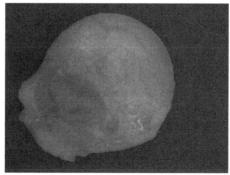

Figure 1: PDC implant 3 weeks after subcutaneous implantation in the region directly in front of the regio interscapularis of a rat

On the other hand, the PCL/BA implant was highly moveable still 5 weeks after implantation and could be easily separated from the subcutis without any signs of adherence. Histological examinations confirmed that the PCL/BA implant was not integrated into the surrounding tissue. Instead of this a clear gap formed between the polymer and the surrounding tissue (fig. 2).

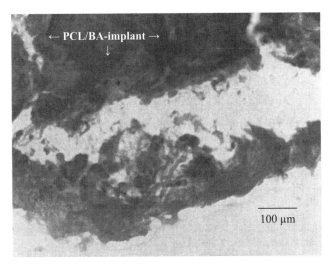

Figure 2: Capsule formation around a PCL/BA-disc 5 weeks after subcutaneous implantation in the region between the shoulders of a rat; collagen fibers were stained blue with Carstairs staining [12].

The periimplantary tissue directly at the PDC- and at the PCL/BA-tissue-interface as well was characterized by a mainly collagen-based connective tissue forming a fibrous capsule around both polymers. The capsules covering the PDC- and the PCL/BA-implants could be distinguished on the basis of different contents on collagen and on the basis of the capsule thickness.

Using polarized light microscopy, it was obvious that the capsule covering the PCL/BA implant was characterized by a higher content ob collagen than the capsule formed around the PDC disc (figure 3).

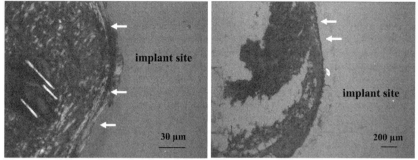

Figure 3: Collagen fibers (bright red) in the fibrous capsule formed around the polymers 3 weeks after implantation; left: PDC, primary magnification 1:200, right: PCL/BA; HE staining; polarized light modus.

Additionally at the whole time period of implantation the capsule around the PCL/BA implant was significantly thicker than the capsule found around the PDC disc ($p < 0.05$, table 1). Thus as positive correlation between capsule thickness and polymer-tissue connection was found. The connective tissue forming the thick capsule covering PCL/BA was not connected to the material in contrast to the strong tissue-polymer connection which was found between PDC and the surrounding capsule like condensed connective tissue. The thickness of the capsule covering the PDC implant stayed stable within a range of 21-35 µm without any significant change after implantation ($p = 0.1625$), while the thickness of the capsule covering the PCL/BA implant significantly increased (approximately 16 fold), even up to mm range, during this monitoring period ($p < 0.0001$).

Table 1: Thickness of the fibrous capsule [µm] formed around the implants (PDC, PCL/BA) after subcutaneous implantation; results are expressed as means ± standard deviation, n = 75 (3 animals per group, 5 slides per sample, and 5 fields of view per sample analyzed).

Implantation time period [weeks]	Thickness of the fibrous capsule formed around the polymer [µm]		p
	PDC	PCL/BA	
1	26 ± 11	40 ± 7	0.012
2	35 ± 9	205 ± 32	< 0.0001
3	21 ± 14	292 ± 112	< 0.0001
5	29 ± 9	656 ± 114	< 0.0001
p	0.1625	< 0.0001	

Figure 1 shows the change of pH directly at the polymer-tissue interface measured after the implantation (1-5 weeks). Directly at the polymer-tissue interface of both copolymers the pH was alkaline over the whole time period of implantation. Significant pH elevation at the PCL/BA-tissue interface was observed at 2 and 5 weeks after the implantation compared to the PDC-tissue interface.

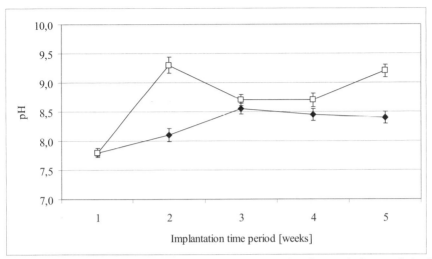

Figure 1: Correlation of the pH at the polymer-tissue interface and time period of subcutaneous implantation (1-5 weeks); physically cross-linked multiblock-copolymers PDC (♦), covalently cross-linked AB polymer networks PCL/BA (□); means ± standard deviation, n = 3.

DISCUSSION

In this *in vivo* study, both physically and covalently cross-linked copolymer based implants were subcutaneously implanted directly in front of the regio interscapularis of rats. This region was used for implantation because of its limited possibilities to get influenced by the animal via scratching and ruzzling. The subcutaneous tissue at the implant site was characterized by an accumulation of collagen in the region nearby the polymer tissue-interface. This collagen enriched subcutaneous tissue formed a fibrous capsule covering both polymers. The capsule formation can be interpreted as a demarcation process of the immunocompetent organism to the copolymer implant. However, in contrast to the capsule covering the PCL/BA disc the capsule covering the PDC implant did not prevent a strong interaction of the subcutaneous cells with this copolymer resulting in a strong integration of PDC into the periimplantary formed connective tissue.

The thickness of the capsules and their collagen content were strongly dependent on the copolymer composition. The capsule covering the PCL/BA implant was more than two times thicker than the capsule covering the PDC copolymer disc even only one week after the implantation. Furthermore, the thickness of the capsule surrounding the PCL/BA constantly increased, even up to mm range after 5 weeks, while thickness of the capsule around the PDC ranged constantly between 21-35 μm. The capsule thickness was inversely correlated to the collagen content of the capsules. The histological examinations showed that the thick capsule which formed around the PCL/BA copolymer had a lower content of collagen fibers than the thin capsule which covered the PDC copolymer. Directly after explantation the capsule covering the PCL/BA disc was mechanically less stable (possibly due to the low collagen content in this capsule) so that it got easily ruptured.

The lower collagen content of the thick capsule formed around the PCL/BA might be related to the pH at the polymer-tissue interface. Kulkarni et al. recently showed that the degradation

147

of PCL as one main component of both copolymers is enzymatically and hydrolytically driven [18]. The hydrolytic degradation of PCL results in hydroxyl caproate by cleavage of ester bondings which can lead to an increase of the pH in an aqueous environment. Considering the fact that a pH range of 7-8 is optimal for collagen secreting cells [3], it might be the more alkaline environment at the PCL/BA-tissue that caused partial inhibition of collagen synthesis and cross-linking. However from the chemical side of view there is no explanation for the more alkaline pH at the PCL/BA-tissue interface compared to the PDC-tissue interface. This has to be reflected in further studies.

CONCLUSION

Although a previous study reported that PCL/BA did not induce synthesis of acute-phase inflammation proteins [10], physically cross linked polymer PDC revealed a mechanically stable interaction between the copolymer and the periimplantary tissue. The results of this study warrant further studies and suggest a potential of PDC as a biomaterial which is intended to be used as implant material in regenerative medicine.

This study was supported by a grant from the German Ministry of Research and Education (BMBF), BioFuture 0311867.

REFERENCES

[1]   Bakker D, van Blitterswijk CA, Hesseling SC, Daems ThW, Grote JJ. Tissue/biomaterial interface characteristics of four elastomers. A transmission electron microscopical study. J. Biomed Mater Res 1990; 24: 277 – 293.

[2]   Eppley BL, Reilly M. Degradation characteristics off PLLA-PGA bone fixation devices. J Craniofac Surg 1997; 8: 116 – 120.

[3]   Gries G, Lindner J. Untersuchung über den Kollagenabbau bei akuten Entzündungen. Z Rheumaforsch 1961; 20: 122.

[4]   Korpela A, Aarnio P, Sariola H, Törmälä P, Harjula A. Bioabsorbable self-inforced poly-L-lactide, metallic and silicone stents in the management of experimental tracheal stenosis. Chest 1999; 115: 490 – 495.

[5]   Lendlein A, Langer B. Biodegradable, elastic shape-memory polymers for potential biomedical applications. Science 2002; 296: 1673 – 1676.

[6]   Lochbihler H, Hoelzl J, Dietz HG. Tissue compatibility and biodegradation of new absorbable stents for tracheal stabilization: an experimental study. J Pediatr Surg 1997; 32: 717 – 720.

[7]   Rickert D, Lendlein A, Kelch S, Franke RP. The importance of angiogenesis in the interaction between polymeric biomaterials and surrounding tissue. Clin Hemorheol Microcirc 2003; 28: 175 - 181

[8]   Rickert D, Lendlein A, Kelch S, Franke RP, Moses MA. Cell proliferation and cellular activity of primary cell cultures of the oral cavity after cell seeding on the surface of adegradable thermoplastic block copolymer. Biomed Technik 2005; 50: 92 – 99.

[9]   Rickert D, Lendlein A, Peters I, Moses MA, Franke RP. Biocompatibility testing of novel multifunctional polymeric biomaterials for tissue engineering applications in head and neck surgery. Eur Arch Otorhinolaryngol 2006; 263: 215 – 222.

[10]   Rickert D, Scheithauer MO, Coskun C, Kelch S, Lendlein A, Franke RP. The influence of a multifunctional, polymeric biomaterial on the concentration of acute phase proteins in an animal model. Clin Hemorheol Microcirc 2007; 36: 301 – 311.

[11]   Robey TC, Välimaa MS, Murphy HS, Törmälä P, Mooney DJ, Weatherly RA. Use of internal bioabsorbable PLGA "finger-type" stents in a rabbit tracheal reconstruction model. Arch Otolaryngol Head Neck Surg 2000; 126: 985 – 991.

[12] Böck P. Romeis. Mikroskopische Technik. Urban & Schwarzenberg, München, 1989

[13] Schauwecker HH, Gerlach H, Planck H, Bücherl. Isoelastic polyurethane prosthesis for segmental trachea replacement in beagle dogs. Artif Organs 1989; 13: 216 – 218.

[14] Tang LP, Eaton JW. Inflammatory responses to biomaterials. Am J Clin Pathol 1995; 103: 466 – 471.

[15] Van Blitterswijk C. Tissue Engineering. Elsevier, Amsterdam. 2008

[16] Van Luyn MJA, Khouw MSL, van Wachem PB, Blaauw EH, Werkmeister. Modulation of the tissue reaction to biomaterials. II. The function of T cells in the inflammatory reaction to crosslinked collagen implanted in T-cell-deficient rats. J Biomed Mater Res 1998; 39: 398 – 406.

[17] Williams DF. On the mechanisms of biocompatibility. Biomaterials 2008; 29: 2941 – 2955.

[18] A. Kulkarni, J. Reiche, J. Hartmann, K. Kratz, A. Lendlein, European Journal of Pharmaceutics and Biopharmaceutics 2008; 68: 46 – 56.

Mater. Res. Soc. Symp. Proc. Vol. 1140 © 2009 Materials Research Society   1140-HH06-33-DD03-33

Development and Characterisation of a Novel Elastin Hydrogel

Nasim Annabi [1], Suzanne M. Mithieux[2], Anthony S. Weiss[2], Fariba Dehghani[1]
[1]School of Chemical and Biomolecular Engineering, University of Sydney, Sydney, 2006, Australia.
[2]School of Molecular and Microbial Biosciences, University of Sydney, Sydney, 2006, Australia.

ABSTRACT

Elastin-based biomaterials offer unique promise as hydrogels for tissue engineering applications. In this study an elastin-based hydrogel was synthesized through coacervation followed by crosslinking under high pressure CO_2. The physical properties of fabricated hydrogel including swelling ratio, pore size, pore interconnectivity, and mechanical properties were tailored by processing pressure. The crosslinked hydrogels fabricated using high pressure CO_2 exhibited superior properties compared with those produced at atmospheric pressure. Dense gas CO_2 strengthened the hydrogel, whereas the gels produced at atmospheric condition were very fragile. High pressure CO_2 produced a highly interconnected porous hydrogel; which resembled the natural elastin structures within the body. Using dense gas CO_2, large channels were induced within the structures of the α-elastin hydrogels. The presence of these channels allowed for the fibroblast cells to penetrate and grow into the 3D structures of fabricated hydrogels.

INTRODUCTION

Hydrogels are highly attractive for tissue engineering applications due to their hydrophilicity and high permeability to oxygen and nutrients [1, 2]. Elastin and elastin-like polypeptides (ELPs) are a unique class of biopolymeric hydrogels due to their capacity for self assembly and phase transition behavior [3, 4]. Elastin-based hydrogels are fabricated from soluble forms of elastin including α-elastin, an oxalic acid-solubilised derivative of elastin, or tropoelastin, soluble precursor of elastin, through two steps: coacervation and crosslinking [5]. The tropoelastin or α-elastin molecules first coacervate in an aqueous solution by intermolecular hydrophobic associations [6]. The molecules are then chemically crosslinked to increase mechanical integrity and form hydrogels. The issues associated with current hydrogel fabrication include the extensive processing time, small pores which cause the lack of cellular growth in 3D structures, the use of toxic solvent, and low mechanical properties. In this study, the feasibility of using dense gas technology to fabricate elastin hydrogel in an aqueous solution was assessed. A dense gas is a fluid at above its critical temperature and pressure with superior properties compared to liquids and gases; higher density than gases and larger mass transfer properties than liquids [7]. Dense gases are excellent fluids for extraction, purification, and a media for conducting reaction [7]. In our study, dense gas carbon dioxide (CO_2) was used to facilitate coacervation and introduce porosity into the hydrogel matrix.

EXPERIMENT

α-elastin extracted from bovine ligament was obtained from Elastin Products Co. (Missouri USA). α-elastin was dissolved in PBS (Phosphate-buffered saline) (10 mM sodium phosphate pH 7.4, 1.35 M NaCl). Glutaraldehyde (GA) was purchased from Sigma. Food grade carbon dioxide (99.99 % purity) was supplied by BOC. GM3348 cell line was obtained from the Coriell Cell Repository. Cells were maintained in Dulbecco's Modified Eagle's Medium (DMEM) supplemented with 10% v/v fetal bovine serum (FBS), penicillin and streptomycin. All tissue culture reagents were obtained from Sigma.

The procedures for the coacervation and synthesis of crosslinked hydrogel under high pressure CO_2 have been described previously in detail [8, 9]. In summary, for coacervation study at high pressure, α-elastin dissolved in PBS was injected into a glass tube located inside the high pressure vessel and the coacervation behavior was monitored visually at each pressure. While the pressure was kept constant, the temperature was increased from 7°C to 37°C and the temperature at which the solution became turbid was monitored. The system was then depressurised slowly and the sample was collected for characterisations. The synthesis of crosslinked hydrogel was conducted at 37°C and various pressures for a period of 30 minutes. In each experiment, the desired amount of crosslinker was added to the elastin solution. The pore characteristics of the hydrogel were governed by the rate of depressurisation. Therefore, the system was depressurised in a controlled manner and the sample was collected after the system was completely depressurised. The properties of the hydrogels fabricated at high pressure CO_2 were compared with those produced at atmospheric conditions at 37°C for 24 hours.

DISCUSSION

Coacervation behavior of α-elastin at high pressure CO_2

The results obtained from coacervation study of α-elastin solution demonstrated that high pressure CO_2 did not disrupt α-elastin coacervation. However, the coacervation temperature was decreased, predominantly due to the pH depression of the aqueous solution and the interaction between CO_2 and hydrophobic domains of α-elastin [8]. It was also found that carbon dioxide at high pressure had no detrimental effect on the secondary structure of α-elastin as confirmed by circular dichroism (CD) analysis. Following exposure to high pressure CO_2 an α-elastin solution was maintained in a coacervated state for a longer time due to the interaction between protein and CO_2 [8]. These results led us to use high pressure CO_2 to fabricate α-elastin hydrogels.

Effect of high pressure CO_2 on crosslinking of α-elastin

The macrostructures of α-elastin hydrogels produced at high pressure CO_2 and atmospheric conditions are shown in Figure 1. The hydrogels produced at high pressure CO_2 were rigid and easily handled. However, the gels produced at atmospheric pressure were very fragile and could not keep the shape in swollen state. This may be due to an insufficient degree of crosslinking throughout the hydrogel matrices produced at atmospheric conditions. High pressure CO_2 facilitated the coacervation of α-elastin and increased the degree of crosslinking through the hydrogel matrix.

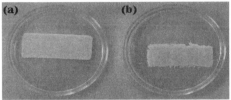

Figure 1. GA cross-linked α-elastin hydrogels produced at (a) high pressure CO_2, and (b) atmospheric pressure

α-elastin hydrogels fabricated in this study exhibited high swelling ratios, in the range of 21-35 g H_2O/ g protein, which was considerably greater or comparable with other elastin-based hydrogels crosslinked by using various cross-linkers [10-12]. As indicated in Figure 2, the swelling ratio of hydrogels enhanced by increasing the processing pressure from 30 bar to 150 bar. The swelling ratio of the hydrogels in water was also greater than those swelled in PBS at 4°C.

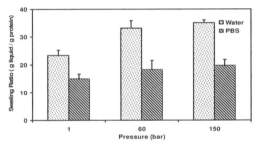

Figure 2. The effect of media on swelling ratio of α-elastin hydrogels at 4 °C (0.5% (v/v) GA)

The pore size and the thickness of the walls between pores were a function of processing pressure. As shown in Figure 3a, thin-walled pores (0.5 μm) with average size of 5 μm were formed by using dense gas CO_2. The highly interconnected structure of elastin hydrogel produced at high pressure CO_2 resembled the natural elastin in the body and allowed for rapid nutrient and oxygen transfer. However, thick-walled pores (4 μm) with average size of 14.3 μm were formed at atmospheric pressure (Figure 3b). The porous structure of hydrogels produced at high pressure CO_2 was induced through the formation of numerous CO_2 bubbles (polymer lean phase) in the polymer rich phase. As a result, the polymer rich phase was distributed over a large interfacial area with a simultaneous thinning of the film of the continuous phase during the cross-linking [9].

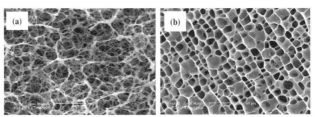

Figure 3. SEM images of α-elastin hydrogel fabricated at (a) high pressure CO_2, and (b) atmospheric pressure

One of the unique features of the dense gas process was the fabrication of large channels within the 3D structures of hydrogels as shown in Figure 4. These channels can promote cellular penetration and growth throughout the matrices. As indicated in Figure, the size of the pores for the hydrogels fabricated at both atmospheric and high pressure CO_2 were smaller than 15 μm. These pores are not suitable for cells penetration and growth; however, they could help for the diffusion of nutrients, oxygen, and waste from cells. Therefore, the cells may grow only on the top surfaces of hydrogel. However, the presence of large channels on the top surface (4a) and also cross-section (4b) of the hydrogel fabricated at high pressure CO_2 may allow for cell infiltration and proliferation into the 3D structures.

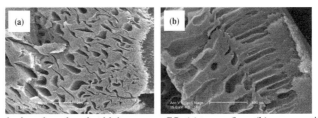

Figure 4: Elastin hydrogel produced at high pressure CO_2 (a) top surface, (b) cross-section

Mechanical properties of fabricated α-elastin hydrogels

Uniaxial compression tests were performed at 60 % strain level on the hydrogels fabricated at high pressure CO_2 in hydrated state, in PBS. The compressive mechanical properties of the sample produced at atmospheric conditions could not be obtained as the PBS-swollen constructs were very fragile. Cyclic stress-strain data for the sample produced by dense gas CO_2 is shown in Figure 5. The compression modulus (slope of the strain-stress curve) of the hydrogel fabricated at high pressure CO_2 was 1.95 ± 0.1 KPa. High pressure CO_2 increased the mechanical integrity of the hydrogel; however, the enhancement was not adequate for the *in vivo* applications of hydrogels. The low compressive mechanical properties of the hydrogels were due to limited GA crosslinking as a result of the low number of lysine residues (less than 1%) in α-elastin that were available for crosslinking. The mechanical properties of fabricated hydrogels can be increased by using another type of crosslinker such as hexamethylene diisocyanate (HMDI) that can react with other amino acids available in α-elastin or through the addition of tropoelastin with high levels of available lysine residues. Research is underway to increase the mechanical properties of fabricated hydrogels.

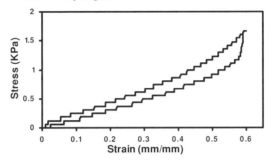

Figure 5: Stress-Strain behavior of an α-elastin hydrogel produced at high pressure CO_2

Attachment and proliferation of fibroblast cell on crosslinked α-elastin hydrogels

Cellular growth and proliferation into the fabricated α-elastin hydrogels was examined by light microscopy analysis. The light microscopy images of adherent fibroblast cells cultured on hydrogels produced at 60 bar CO_2 and atmospheric pressure are shown in Figure 6. Haematoxylin and eosin was used for staining, as a result the cells appear as dark gray and the α-elastin scaffolds as light gray. As shown in Figure 6a, fibroblast cells were able to grow into the 3D structures of α-elastin due to the presence of large chancels induced by high pressure CO_2. However, as indicated in Figure 6b, cells were developed as a monolayer on the surface of the hydrogel fabricated at atmospheric conditions due to the presence of small pores on the surfaces. The cell proliferation into the 3D structure of hydrogels produced by dense gas CO_2 was also confirmed by SEM analysis on cell seeded hydrogels [9].

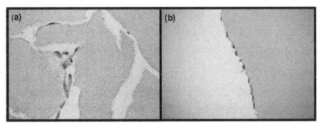

Figure 6. Haematoxylin and eosin staining of fibroblast cells cultured on hydrogel produced at (a) high pressure CO_2, and (b) atmospheric pressure.

CONCLUSIONS

This study demonstrated the feasibility of using dense gas CO_2 to fabricate elastin-based hydrogels in an aqueous solution. It was found that high pressure CO_2 did not disrupt the coacervation and secondary structure of α-elastin. High pressure CO_2 strengthened the hydrogel and produced hydrogels with highly interconnected thin-walled pores that resembled the natural elastin. Generation of large channels into the 3D structure of hydrogel was a unique feature of the dense gas process. The presence of these large channels facilitated the cellular penetration and growth into the hydrogel matrix. Further research is required to increase the mechanical properties of fabricated hydrogels for tissue engineering applications.

REFERENCES

1. N. A. Peppas, J. Z. Hilt, A. Khademhosseini, and R. Langer," *Adv. Mater.***18**, 1345-1360 (2006).
2. A. S. Hoffman," *Adv Drug Deliv Rev* **54**, 3-12 (2002).
3. A. J. Simnick, D. W. Lim, D. Chow, and A. Chilkoti, *Polym Rev* **47**, 121-154 (2007).
4. W. F. Daamen, J. H. Veerkamp, J. C. M. van Hest, and T. H. van Kuppevelt, *Biomaterials* **28**, 4378-4398 (2007).
5. S. M. Mithieux and A. S. Weiss, *Advances in Protein Chemistry* **70**, 437-461 (2005).
6. B. Vrhovski, S. Jensen, and A. S. Weiss, *Eur J Biochem* **250**, 92-8 (1997).
7. R. A Quirk, R. M. France, K. M Shakesheff, S. M. Howdle, *Curr Opin in Solid State &Mater Sci* **8**, 313-321 (2005).
8. F. Dehghani, N. Annabi, P. Valtchev, S. M. Mithieux, A. S. Weiss, S. G. Kazarian, and F. H. Tay, *Biomacromolecule* **9**, 1100-1105 (2008).
9. N. Annabi, S. M. Mithieux, A. S. Weiss, and F. Dehghani, *Biomaterials* **30**, 1-7 (2009).
10. D. W. Lim, D. L. Nettles, L. A. Setton, and A. Chilkoti, *Biomacromolecule* **8**, 1463-1470 (2007).
11. P. J. Nowatzki and D. A. Tirrell, *Biomaterials* **25**, 1261-1267 (2003).
12. J. B. Leach, J. B. Wolinsky, P. J. Stone, and J. Y. Wong, *Acta Biomater***1**, 155-64 (2005).

Mater. Res. Soc. Symp. Proc. Vol. 1140 © 2009 Materials Research Society   1140-HH06-03-DD03-03

Skin-external device integration using porous cationic poly(DMAA-co-AMTAC) hydrogels

Antonio Peramo[1], Joong Hwan Bahng[2], Cynthia L. Marcelo[3], Nicholas Kotov[2] and David C. Martin[1]
[1]Department of Materials Science and Engineering
[2]Department of Chemical Engineering
[3]Department of Surgery
Ann Arbor, MI 48109, U.S.A.

ABSTRACT

Porous poly(DMAA-co-AMTAC) hydrogels (N,N-Dimethylacrylamide (DMAA) copolymerized with (3-Acrylamidopropyl)-trimethylammonium chloride (AMTAC)), fabricated using the inverted colloid crystal method, were integrated with living human skin. Full thickness human breast skin explants discarded from surgeries were cultured for five days at the air-liquid interface using a Transwell culture system. Cylindrical, disks or other shaped hydrogels were placed inside the skin explants fitting the punctures produced by punch biopsies. Full section histological analysis of the skin explants with the inserted hydrogel was then performed. Results indicate that poly(DMAA-co-AMTAC) hydrogels induce substantial extracellular matrix material deposition, maintain excellent dermal integrity in the contact areas with the skin and induce dermal fibers to integrate into the 190 μm hydrogel pores.

INTRODUCTION

There has been a steady increase in the number of medical procedures with permanently implanted transcutaneous devices, which range from glucose sensors and catheters to more complex biointegrated prosthetics. A typical complication is the break-down of the skin around the device, but the most prevalent problem producing device failure arise from infection processes [1-3]. These devices show high variability in their intrinsic properties, with a disparate number of material composition, surface structure, porosities and topologies [4]. This variability hampers the investigation of a general solution for this unresolved problem.

For this application, we were seeking to use a soft, biocompatible material inducing dermal integration that could be produced as hydrogels with controlled porosities. We have been investigating the use of N,N-Dimethylacrylamide (DMAA) copolymerized with (3-Acrylamidopropyl)-trimethylammonium chloride (AMTAC) (poly(DMAA-co-AMTAC)) for its use in the integration of human skin with percutaneous devices. This material has the interesting property that induces cell adhesion without the need for surface modification of the hydrogel [5-6].

EXPERIMENT

Skin preparation

Full thickness human breast skin explants from discarded material from surgeries performed at the University of Michigan Health System were used. After removal of subcutaneous fat, the tissue was rinsed abundantly with PBS 1X containing 125 μg/ml of Gentamicin (Invitrogen/GIBCO, Carlsbad, CA) and 1.87 μg/ml of Amphothericin B (Sigma-Aldrich, Milwaukee, WI) and placed in aliquots of the same medium in an incubator at 37^0C for 2 hours with change of medium every hour. The culture medium used was EpiLife (Cascade Biologics, Portland, OR). In addition, the medium was supplemented with 75 μg/ml of Gentamicin and 1.125 μg/ml of Amphothericin B. The final concentration of calcium used in the culture medium was 1.2mM.

Preparation of 3D Inverted Colloidal Crystal (ICC) poly(DMMA-co-AMTAC) hydrogels

The following materials were used for the hydrogel synthesis: the neutral monomer N,N-Dimethylacrylamide ($CH_2=CHCON(CH_3)_2$, DMAA, Aldrich), the cationic monomer (3-acrylamidopropyl)-trimethylammonium chloride ($H_2C=CHCONH(CH_2)_3N(CH_3)_3Cl$, AMTAC, Aldrich), the cross-linker N,N'-methylenebisacrylamide (($CH_2=CHCONH)_2CH_2$, NMBA, Sigma), the free radical initiator potassium persulfate (K_2SO_4, KPS, Sigma). De-ionized distilled (DDI) water (E-pure, Barnstead) was used to make the pre-polymer solution and the free radical solution. An aqueous suspension of polystyrene (PS) microspheres (Duke Scientific, 3 x 10^4 particles per ml with 1.4% size distribution) was used for the construction of the colloidal crystal. The poly(DMAA-co-AMTAC) was prepared from co-polymerization between DMAA and AMTAC in an aqueous environment. The monomer chains were chemically cross-linked by NMBA, and the polymerization was triggered by the free radical initiator KPS. Colloidal crystals (CC) were used as an invertible mold for the ICC poly(DMAA-co-AMTAC) hydrogel scaffolds. The construction of CC followed Lee's reported method with only minor changes [8]. The only component comprising its structure is non-cross-linked polystyrene microspheres. The resulting CC was immersed in the oxygen free gel precursor solution. The gel precursor was infiltrated in between the microspheres via centrifugation at 3800 RPM for 15 minutes. Again in an oxygen free environment, aqueous KPS solution (3 w/v%) was added at a ratio of 1:10 by volume and then re-centrifuged for a further 5 minutes to allow thorough penetration into the crystal. Subsequent heat treatment at 75°C for 3 hours and 60°C overnight completed the polymerization into a poly(DMAA-co-AMTAC) hydrogel. After polymerization, excess hydrogel pieces were removed until the surface of microspheres were exposed. The hydrogel encapsulated CC was then immersed in a THF bath for 48 hours to dissolve away the internal PS microspheres resulting in a 3D ICC polyelectrolyte hydrogel tissue culture scaffold. The THF bath was renewed after 24 hours. The average pore size of the hydrogels was 190 μm.

Culture of poly(DMMA-co-AMTAC) hydrogels with skin specimens

After preparation of skin, skin specimens of approximately 1.5 cm^2 were cut using a scalpel and cultured for 5 or 10 days at 37^0 C and 5% CO_2 atmosphere, epidermal side up at the air-liquid interface, in a Transwell system consisting of 6-well Transwell carriers (Organogenesis, Canton, MA) and six Corning Costar supports (Fisher Scientific, Pittsburgh, PA) [7]. Organotypic cultures of human skin can be maintained up to two weeks, thus the decision to establish time-points at earlier dates, in order to work with better preserved specimens. Culture medium was changed every 24 hours and the stratum corneum remained constantly exposed to the air. For histological analysis, specimens with hydrogels were dehydrated in a graded series of ethanol, infiltrated and embedded in JB-4 embedding resin

(Electron Microscopy Sciences, Hatfield, PA). The blocks were sectioned at a thickness of 3 microns and the sections were stained with either toluidine blue or a mixture of toluidine blue and basic fuchsin, as shown in Figure 3. Figure 1 shows the preparation of the cultured skin with the poly(DMAA-co-AMTAC) hydrogels. The specimens were punctured with a scalpel and the hydrogels were placed inside the puncture. The hydrogels were made sterile by UV overnight irradiation before use. The hydrogels were washed first with PBS and then with culture medium before insertion in the punctures. Figure 1B indicates the method used for obtaining the tissue sections for histological analysis, sectioning the specimens in half, tracing a line across the center of the perforation. In general, the hydrogels remained tightly in contact with the specimens. Figure 1C show details of the specimens with the hydrogel placed in the perforation and the Transwell culture system used.

Figure 1. The hydrogels were inserted as shown in A) and after the culture period were sectioned as in B), that is, the specimen and hydrogel were sectioned in half and then SEM imaged. The skin was cultured using a Transwell system for culturing organotypic tissue cultures, C).

DISCUSSION

In this work, poly(DMAA-co-AMTAC) hydrogels were evaluated for their possible use to facilitate integration with human skin, and experiments using full human skin were performed. Three dimensional matrices are generally better suited for tissue culture and integration [8], thus the use of the copolymer in form of hydrogel. The purpose of the hydrogel is to provide a surface for cell adhesion and also to maintain mechanical stability at the injury site [9]. The monomer used to construct the hydrogel (N,N-Dimethylacrylamide (DMAA)) has been proven to polymerize into a highly hydrophilic and biocompatible hydrogel [10] and has been used for contact lenses applications. To impart a cationic nature to the hydrogel, (3-Acrylamidopropyl)-trimethylammonium chloride (AMTAC) was copolymerized with DMAA.

We used organotypic cultures [11] of human skin explants because they represent an attractive model to use in the study of the biomaterial-skin interface [12]. In accordance with previous studies of organotypic cultures of human skin explants [13], no substantial morphological changes of the skin specimens were observed during the initial five days for control specimens.

Figures 2A and 2B are SEM pictures of the specimen pictured in Figure 1B that provided initial evidence of the strong affinity of the dermis with the hydrogels. As observable in Figure 2B, several contact points between hydrogel and dermis remained even after shrinkage due to the vacuum process for SEM imaging.

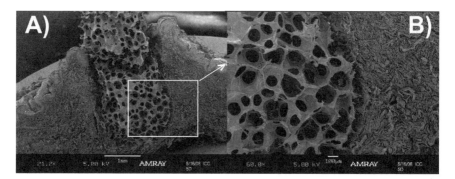

Figure 2. SEM images of a skin specimen cultured for five days with an inserted poly(DMAA-co-AMTAC) hydrogel.

Based on this initial evidence histological analysis of the samples and more extended culture periods (10 days) were performed to more accurately determine the effectiveness of the copolymer to induce skin integration. Histological features of the specimen pictured in Figure 2 are shown in Figure 3. The sections stained with toluidine blue (Figure 3A and 3B) indicate that the tissues maintained excellent dermal integrity in the contact areas with the skin and induced dermal fibers to integrate and move into the hydrogel pores. In contrast to two dimensional cultures where newly formed ECM is difficult to observe [14], dermal integration was detected far away from the dermal contact areas, as seen in Figure 3B (indicated by white arrows).

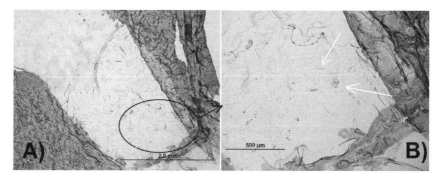

Figure 3. Toluidine blue staining of sections of a skin specimen with hydrogel. Deposits of extracellular matrix materials are readily visible in B).

More extended culture periods (10 days) were performed in order to observe increased depositions of the extracellular matrix materials. This can be observed in more detail in Figure 4 where an SEM image of a skin specimen was cultured with a hydrogel for ten days. The thickness of the deposited layer is about 1.5 μm.

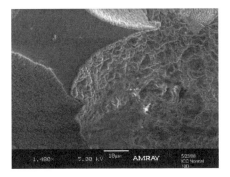

Figure 4. SEM image of hydrogel after ten days of culture with skin specimen showing a thick (~ 2 μm) layer of deposited extracellular matrix materials.

CONCLUSIONS

The results presented in this manuscript strongly indicate that poly(DMAA-co-AMTAC) hydrogels induce substantial extracellular matrix materials deposition by the skin specimens surrounding them and maintain excellent dermal integrity in the contact areas with the skin. Our results suggest that this type of soft, biodegradable material could be used as a general interface that induces skin integration with percutaneous devices in contact with skin.

ACKNOWLEDGMENTS

We thank the Microscopy & Image-Analysis Laboratory (MIL) of the University of Michigan School of Medicine for their work on specimen preparation and Ming Qin, from the Department of Chemical Engineering, University of Michigan, Ann Arbor, for his help with the spin coater. This report is presented as part of the research efforts within the Army Research Office Multidisciplinary University Research Initiative award on Bio-Integrating Structural and Neural Prosthetic Materials (contract number W911NF-06-1-0218, proposal number 50376-LS-MURI) and we gratefully acknowledge the funding provided.

REFERENCES

1. G.D. Winter, *J. Biomed. Mater. Res. Symp. No. 5 (Part I)* **101**, 99 (1974).

2. J. Mahan, D. Seligson, S.L. Henry, P. Hynes and J. Dobbins, *Orthopedics* **14**, 305 (1991).

3. A.F. Von Recum, *J. Biomed. Mater. Res.* **18**, 323 (1984).

4. C.J. Pendergrass, A.E. Goodship and G.W. Blunn, *Biomaterials* **27**, 4183 (2006).

5. F. Rosso, A. Barbarisi, M. Barbarisi, O. Petillo, S. Margarucci, A. Calarlo and G. Peluso, *Mater. Sci. Eng. C-Biomimetic and Supramolecular Systems* **23**, 371 (2003).

6. H.S. Kang, S.H. Park, Y.G. Lee and T.I. Son, *J. App. Polym. Sci.* **103**, 386 (2007).

7. M. Bedoni, C. Sforza, C. Dolci and E. Donetti, *J. Dermatol. Sci.* **46**, 139 (2007).

8. J. Lee, M.J. Cuddihy and N.A. Kotov, *Tissue Eng. Part B Rev*. **14**, 61 (2008).

9. D.W. Hutmacher, *Biomaterials* **20**, 2529 (2000).

10. M.P. Mullarney, T.A.P. Seery and R.A. Weiss, *Polymer*, **47**, 3845 (2006).

11. J.H. Resau, K. Sakamoto, J.R. Cottrell, E.A. Hudson and S.J. Meltzer, *Cytotechnology* **7**, 137 (1991).

12. I.C. Le Poole, R.M. Van der Wijngaard, W. Westeerhof, J.A. Dormans, F.M. Van der Berg, R.P. Verkruisen, K.P. Dingemans and P.K. Das, *Pigment. Cell. Res.* **7**, 33 (1994).

13. R. Tammi, C.T. Jansen and R. Santti, *J. Invest. Dermatol.* **73**, 138 (1979).

14. C.J. Doillon, F.H. Silver and R.A. Berg, *Biomaterials* **8**, 195 (1987).

Mater. Res. Soc. Symp. Proc. Vol. 1140 © 2009 Materials Research Society   1140-HH09-01

Designing Enzyme-Triggered Hydrogels for Biomedical Applications Using Self-Assembling Octapeptides

Elisabeth Vey[1], Alberto Saiani[2] and Aline F. Miller[2]
[1] Manchester Interdisciplinary Biocentre, The University of Manchester, 131 Princess Street, Manchester, M1 7DN, UK
[2] School of Materials, The University of Manchester, Grosvenor Street, Manchester, M1 7HS, UK

ABSTRACT

We demonstrate that protease can be used to trigger the synthesis of ionic complementary peptides and above a critical peptide concentration this results in a sol-gel transition. Upon addition of the protease the tetra peptide FEFK partially hydrolyses to FE and FK and this subsequently encourages the synthesis of hexa, octa and decapeptide through reverse hydrolysis. The octapeptide is the most favored product probably due to its high self-assembling ability as once formed they self-assemble and become trapped into β-sheet rich nanofibers that subsequently entangle to form a self-supporting, elastic hydrogel. This novel method opens up the possibility of synthesizing a diverse library of ionic peptides that undergo a sol-gel transition with no harsh chemicals and water being the only by-product.

INTRODUCTION

Stimuli-responsive, or smart, materials have received considerable attention recently due to their potential biomaterials applications in for example drug delivery,[1] bio-sensing[2] or regenerative medicine.[3] There are a number of studies in the literature using pH,[4] ionic strength,[5] light[6] or electric/magnetic fields[7] as external stimuli to induce the self-assembly of small molecules into hydrogels for such biomaterials applications. Enzyme catalysis is one further class of stimuli that is known to induce changes in material properties.[8-11] Typically, enzymes have been used to hydrolyze materials, mainly polymers, for controlled drug delivery applications. Recently, however, enzymes have been exploited to trigger molecular assembly via reversed hydrolysis.[12,13] Examples for this latter method exploit a class of enzymes called proteases which are commonly known to hydrolyze peptide bonds. However, it has been shown that proteases can be encouraged to work in reverse when the reaction product is thermodynamically stabilized relative to its precursors. One interesting example involves the coupling of a non-gelling N−(fluorenylmethoxycarbonyl) (Fmoc) amino acid to a dipeptide to create a Fmoc-tripeptide using proteases.[12] In dilute aqueous solutions hydrolysis of the peptide is favored, however, the equilibrium was found to shift in favor of the Fmoc-tripeptide synthesis as a further equilibrium favored the self-assembly of these molecules into nanofibrous structures driven by π-stacking. These fibers eventually became physically entangled and formed a hydrogel. This example involved the use of poorly soluble peptide derivatives resulting in a suspension to gel transition upon addition of the enzyme. Such enzyme responsiveness is a particularly attractive hydrogel fabrication route as it is highly selective, occurs under mild conditions with water being the only by-product.

Here we exploit the reverse hydrolysis method to synthesize ionic peptides (containing alternating charged and non-charged amino acids) that are well known to self-assemble into β-sheet rich fibrillar hydrogels. Such peptides are gaining increasing popularity in the literature due to their potential biomaterials applications and for aiding the understanding of the general paradigms that govern molecular self-assembly.[14-16] The key idea behind this is that we can start with a solution containing readily soluble, short, easy to synthesize peptides, and simply use an enzyme to couple the precursor peptides together which will result in a sol-gel transition and hence a 3-dimensional matrix ready for biomaterials applications. This method opens up the possibility of synthesizing a diverse range of ionic peptides with no harsh chemicals and only water as the by-product. In addition this is a particularly attractive hydrogel fabrication route for 3D tissue engineering applications as it can be used to trigger gelation in the presence of cells, thus incorporating cells throughout the hydrogel scaffold. Based on Zhang's and our previous work with ionic peptides[16,17] we selected the tetrapeptide FEFK (F is phenylalanine, E is glutamic acid and K is lysine), as this peptide forms hydrogels only at very high concentrations (> 300 mg mL^{-1}) while the octapeptide FEFKFEFK forms hydrogels at low concentrations (~ 10 mg mL^{-1}).[17] The enzyme selected here was Thermolysin from Bacillus thermoproteolyticus rokko as it is known to hydrolyze peptide bonds on the amino side of hydrophobic residues,[18] in our case F.

EXPERIMENT

Materials and peptides' synthesis: Amino acids, activator (HCTU) and Wang resin were purchased from Novabiochem (Merck) and used as received. All other reagents and solvents were purchased from Aldrich and used without further purification. FEFK was synthesised via standard solid-phase synthesis on a ChemTech ACT 90 peptide synthesiser (Advance ChemTech Ltd., Cambridgeshire, UK), using Fmoc-Lys(Boc)-Wang resin (mesh = 200, loading = 0.7 mmol g^{-1}); coupling of amino acids was done using standard solid phase synthesis protocols with HCTU as activator; deprotection of side groups was done using piperidine; cleavage was done by using a TFA-anisole mixture. Each reaction step was confirmed by Kaiser test. The peptide was recovered and washed in cold ether and freeze-dried. The peptide purity was estimated to be around 90%.

Reversed Phase - High Performance Liquid Chromatography (RP-HPLC): RP-HPLC analysis were performed at 35 °C on Ultimate 3000 HPLC (Dionex) system using a gradient of HPLC grade water and acetonitrile (ACN) containing 0.1 % triflouroacetic acid (TFA) as eluents. The HPLC system was equipped with a C18 column (3 μm, 4.6 × 150 mm) and a UV detector and the flow rate was 1 mL min^{-1}. The samples were prepared by diluting 20 μL of peptides in a 80/20 mixture water/ACN. 100 μL of the solution was then injected into the column using a ACC-300A autosampler. The data were analysed using Chromoleon CHM-1 software. The normalised relative concentrations of tetra, hexa, octa and decapeptides were calculated taking into account the area of the peaks corresponding to these peptides only and were obtained by dividing the area of the peak corresponding to each peptide by the sum of peaks areas corresponding to the four peptides.

Matrix Assisted Laser Desorption Ionisation – Time of Flight Mass Spectrometer (MALDI-TOF MS): The mass of the samples was determined on a Axima CFR mass spectrometer (Shimadzu Biotech) in reflectron positive mode using a nitrogen laser (337 nm). The matrix, α-cyano-4-hydroxycinnamoe acid, was prepared in 250 μL of CAN, 25 μL of 2 % formic acid and 225 µL of HPLC grade water. 20 µL of sample were diluted in 1 mL of HPLC grade water. 1 µL of sample and 1 µL of matrix solution were mixed onto a stainless steel sample plate and the solvent was removed by evaporation for ~ 10 min. All mass spectra were generated by collecting 100 laser shots and laser strength was adjusted to obtain optimal signal-to-noise ratio. The data were analysed using Kompact 2.4.1 software.

Attenuated total reflectance – Fourier Transform infrared spectroscopy (ATR-FTIR): Secondary structures of the peptides were characterised by ATR-FTIR on a Thermo Nicolet 5700 spectrometer using a Smart multi-Bounce ARK accessory (Thermo Nicolet) with a zinc selenide crystal. A background spectrum, obtained with HPLC grade water, was taken prior to any sample analysis and subtracted from all the sample spectra. Solution of FEFK and thermolysin were mixed and directly placed onto the crystal for analysis. Spectra were taken every 20 min, 64 scans were collected and averaged in order to obtain a good signal-to-noise ratio. The software OMNIC 7.2 was used to acquire and process the data.

Rheology: Rheological studies were undertaken on a stress-controlled Bohlin C-CVO rheometer, equipped with a Peltier device to control the temperature. A parallel plate geometry of 40 mm diameter and a 0.5 mm gap were used; to minimise evaporation, a solvent trap was used. The frequency was fixed to 1 Hz and the strain to 0.01. Solution of FEFK and thermolysin were mixed and directly loaded onto the stage for analysis. Elastic and viscous moduli were measured every 8 s.

Transmission Electron Microscopy (TEM): Optical micrographs TEM experiments were performed on a Tecnai 10 TEM operating at 100 keV. Data were collected onto Kodao SO-163 films. Micrographs were scanned at 1600 dpi using a UMAX 2000 transmission scanner, giving a specimen level increment of 3.66 Å/pixel. Sample gels were diluted 20-fold. A 10 µL of the resulting solution was applied to carbon-coated copper grid (400 mesh, Agar scientific). After blotting on Whatman 50 filter paper, the grid was washed in double distilled water for 30 s and blotted. The sample was then stained with 10 µL of 2 % (w/v) uranyl acetate for a minute for 10 s and examined under the TEM.

RESULTS AND DISCUSSION

FEFK (Figure 1A) solutions with concentration, C, ranging from 20 to 200 mg mL^{-1} were prepared by dissolving the desired quantity of tetrapeptide in distilled water and adjusting the pH to 7 by adding a few drops of a 1 mol L^{-1} NaOH solution. Thermolysin was subsequently added to each solution at a concentration of 0.3 mg mL^{-1}. When C was < 50 mg mL^{-1} no increase in viscosity was observed and the sample remained in a liquid state even after several days of incubation at 25 °C (Figure 1B). Mass spectrometry revealed that the tetrapeptide was no longer present in the sample, instead thermolysin had hydrolyzed the tetrapeptide leading to the formation of the two dipeptides FE and FK. When the FEFK concentration was > 60 mg mL^{-1}, clear, self-supporting gels formed (Figure 1B) at different times depending on the peptide concentration: the higher the concentration, the quicker the gelation. Here we will focus on the detailed characterization of one of these hydrogels, namely the hydrogel formed from 200 mg mL^{-1} of FEFK. At this concentration a strong self-supporting gel is obtained after ~ 30 minutes.

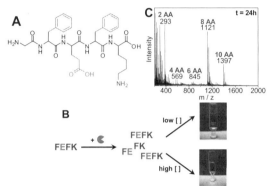

Figure 1. (A) Chemical structure of FEFK. (B) Chemical reaction of FEFK with thermolysin. (C) MALDI-TOF spectrum of FEFK (200 mg mL^{-1}) at t = 24 hours after addition of thermolysin.

The gelation dynamics of the system were followed using oscillatory rheology where the elastic, G', and loss, G", moduli were monitored as a function of time, t (Figure 2A). At early times (t < 8 min) G" is larger that G' as expected for a liquid. As the reaction progressed both moduli started to increase and after 8 mins G' becomes larger than G" suggesting that the materials start to have more of a "solid-like" behavior. G' and G" increase significantly up to 20 mins and then start to level off. After 30 mins G' is found to be two orders of magnitude larger than G" which is characteristic of a "solid-like" material and agrees with our visual observation which revealed the formation of a strong self-supporting hydrogel after 30 mins of incubation. After 40 min, G' was ~ 2500 Pa, which is similar to the value obtained for FEFKFEFK at a concentration of 50 mg mL^{-1}. The point at which the G' and G" curves cross-over (G' = G") is often defined as the "gel point", in this case the cross-over occurs after ~ 8 min.

Mass spectrometry was used to monitor the fate of FEFK at discrete time points after the addition of thermolysin. Several peptide lengths were found to form with time. After only 30 mins a peak at 294 g mol^{-1} appeared which corresponds to the dipeptides FE (294 g mol^{-1}) and FK (293 g mol^{-1}). This indicates that thermolysin has hydrolyzed some of the tetrapeptide. Peaks at 845 and 1121 g mol^{-1} also appeared which correspond to hexa and octapeptides respectively, suggesting that reverse hydrolysis between dipetides and/or tetrapeptides also occurs. After 24 h, a peak for the decapeptide was also found at 1396 g mol^{-1} (Figure 1C). It should be noted that due to the presence of the FE and FK dipetides and the combinatorial nature of the reverse hydrolysis reaction it is probable that a dynamic library of peptide with varied sequences is formed. The molar mass of E and K differ by only 1 unit, therefore mass spectrometry is unable to differentiate between the different sequences possible for each size of peptide.

In order to quantify the normalized relative concentrations of tetra, hexa, octa and decapeptide, HPLC was used (Figure 2B). It is clear that the quantity of tetrapeptide decreased rapidly immediately after addition of the enzyme. Over the first hour of reaction hexa and octapeptides started to form with hexapeptide dominating initially. The presence of decapeptide was detected later, from 2 hours onwards. This peptide will form only once hexapeptides and/or octapeptides have formed. After 10 hours the relative concentration of hexapeptide started to decrease while that of octa and decapeptides continued to increase suggesting the conversion of the hexapeptides into octa and decapeptides. This decrease in quantity of hexapeptide suggests that this peptide is probably weakly integrated into the gel network leaving it accessible for further reactions with time. The relative concentrations of the octa and decapeptides increased steadily and after 70 hours the main product of the reaction was octapeptides.

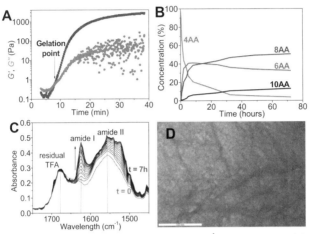

Figure 2. (A) Oscillatory rheology of FEFK (200 mg mL^{-1}) upon addition of thermolysin. (B) Conversion (%) versus time showing the formation of hexa, octa and decapeptides. (C) Infrared spectra of FEFK (200 mg mL^{-1}) at different times after addition of thermolysin. (D) TEM micrograph of enzymatically prepared gel, the scale bar represent 100 nm.

To investigate whether β-sheet rich nanofibers did indeed form, the samples were examined using Fourier transform infra-red spectroscopy (FTIR) and transmission electron microscopy (TEM). FTIR spectra were recorded at 20 min intervals and results for the first 7 hours are given in Figure 2C. It is clear that a band at 1624 cm^{-1}, corresponding to the amide I vibration, appeared and became stronger over time. This band is indicative of the formation of β-sheet rich structure.[19] Samples after 70 hours of reaction were viewed under TEM to detect the presence of any fibrillar structures and a typical micrograph obtained is given in Figure 2D. As can be seen thin fibrils of ~ 4–10 nm diameter and ≥ 1 μm in length are observed. Such dimensions are similar to those formed from the octapeptide.[17]

CONCLUSIONS

In summary, thermolysin triggers the reverse hydrolysis and subsequent gelation of the ionic peptide FEFK. Initially, the protease partially hydrolyses the tetrapeptide into the two dipeptides FE and FK and subsequently encourages the synthesis of hexa, octa and decapeptide through reverse hydrolysis with the octapeptide being the main product after 70 hours. Octapeptide is the most favored product probably due to its high self-assembling ability;[14] as once formed they will self-assemble and become trapped into β-sheet rich nanofibers that subsequently entangle to form a self-supporting hydrogel. Once trapped in this state the peptides are prevented from having any further reaction with the enzyme. This novel method of forming ionic peptides and subsequently hydrogels will undoubtedly have significant impact in the synthesis of longer peptides and will also find important applications in the formation of scaffolds for in vivo tissue engineering or for use in enzyme detection

ACKNOWLEDGMENTS

The authors gratefully acknowledge financial support from The Leverhulme Trust (F/00120/AR) and Richard Collins and Stephen Boothroyd for TEM measurements.

REFERENCES
(1) Hynd, M. R.; Turner, J. N.; Shain, W. J. Biomater. Sci., Polym. Ed. 2007, 18, 1223-1244.
(2) Hersel, U.; Dahmen, C.; Kessler, H. Biomaterials 2003, 24, 4385-4415.
(3) Peppas, N. A.; Sefton, M. V. Advances in Chemical Engineering; Academic Press: San Diego, 2004; Vol. 29.
(4) Jayawarna, V.; Ali, M.; Jowitt, T. A.; Miller, A. E.; Saiani, A.; Gough, J. E.; Ulijn, R. V. Adv. Mater. 2006, 18, 611-614.
(5) Ozbas, B.; Kretsinger, J.; Rajagopal, K.; Schneider, J. P.; Pochan, D. J. Macromolecules 2004, 37, 7331-7337.
(6) Haines, L. A.; Rajagopal, K.; Ozbas, B.; Salick, D. A.; Pochan, D. J.; Schneider, J. P. J. Am. Chem. Soc. 2005, 127, 17025-17029.
(7) Ahn, S. K.; Kasi, R. M.; Kim, S. C.; Sharma, N.; Zhou, Y. X. Soft Matter 2008, 4, 1151-1157.
(8) Ulijn, R. V.; Smith, A. M. Chem. Soc. Rev. 2008, 37, 664-675.
(9) Adler-Abramovich, L.; Perry, R.; Sagi, A., Gazit, E.; Shabat, D. ChemBioChem 2007, 8, 859-862.
(10) Dos Santos, S.; Chandravarkar, A.; Mandal, B.; Mimna, R.; Murat, K.; Saucede, L.; Tella, P.; Tuchscherer, G.; Mutter, M. J. Am. Chem. Soc. 2005, 127, 11888-11889.
(11) Yang, Z. M.; Gu, H. W.; Fu, D. G.; Gao, P.; Lam, J. K.; Xu, B. Adv. Mater. 2004, 16, 1440-1444.
(12) Toledano, S.; Williams, R. J.; Jayawarna, V.; Ulijn, R. V. J. Am. Chem. Soc. 2006, 128, 1070-1071.
(13) Yang, Z. M.; Liang, G. L.; Xu, B. Soft Matter 2007, 3, 515-520.
(14) Caplan, M. R.; Schwartzfarb, E. M.; Zhang, S. G.; Kamm, R. D.; Lauffenburger, D. A. Biomaterials 2002, 23, 219-227.
(15) Schneider, J. P.; Pochan, D. J.; Ozbas, B.; Rajagopal, K.; Pakstis, L.; Kretsinger, J. J. Am. Ceram. Soc. 2002, 124, 15030-15037.
(16) Zhang, S. G.; Holmes, T.; Lockshin, C.; Rich, A. Proc. Natl. Acad. Sci. U. S. A. 1993, 90, 3334-3338.
(17) Mohammed, A.; Miller, A. F.; Saiani, A. Macromol. Symp. 2007, 251, 88-95.
(18) Morihara, K.; Tsuzuki, H. Eur. J. Biochem. 1970, 15, 374-380.
(19) Kong, J.; Yu, S. Acta Biochim. Biophys. Sin. 2007, 39, 549-559.

Mater. Res. Soc. Symp. Proc. Vol. 1140 © 2009 Materials Research Society   1140-HH06-29-DD03-29

Platelet Response on Poly(D,L -lactide-*co*-glycolide) (PLGA) Film with Nano-structured Fillers

Li Buay Koh [1], Isabel Rodriguez [2], Subbu S Venkatraman [1]
[1] School of Materials Science and Engineering, Nanyang Technological University, Singapore, 639798, Singapore
[2] Institute of Materials Research & Engineering, A*STAR (Agency for Science, Technology and Research), Singapore, 117602, Singapore

ABSTRACT
Thrombosis is a frequent complication associated with blood-contacting devices such as catheters and artificial stents. Current control of thrombus formation via use of anticoagulant therapies is clearly not ideal, as complications such as thrombocytopenia, neutropenia and hemorrhage can arise due to their usage. To avoid these problems, biomaterials with higher level of blood compatibility are necessary. Platelet adhesion and activation onto an implant surface are crucial events in the formation of thrombus resulting from the interaction between the flowing blood and the foreign material. Surface chemistry and topography are the parameters that greatly determine the adhesion of platelets to a surface. While a lot of research has been performed on the development of materials, the effect of surface topography on platelet adhesion is relatively unexplored. In this work, we show evidence that a nano structured polymer surface significantly reduces platelet adhesion as compared to pristine films. Nano-structured fillers were prepared on poly(D,L -lactide-*co*-glycolide) (PLGA) films by infiltrating a PLGA solution into a nano porous anodized alumina (NPAA) template. The aspect ratio of the nano-sized fillers proved to be one of the important parameter influencing the amount of platelet adhesion. The results indicate that nanotopographic modifications of surfaces can elicit desired interfacial platelet response which could be significant in the development for new polymeric blood-contacting materials with low thrombogenicity.

INTRODUCTION
Thrombosis is the formation of a blood clot inside a blood vessel, which obstructs the flow of blood through the circulatory system. This natural phenomena occurs when a blood vessel is injured or comes into contact with a foreign material [1]. Biomaterials employed in the construction of blood-contacting medical devices can also cause the generation of thrombus. Blood compatibility is one of the key problems of biomaterials today and thrombosis and thromboembolism are important clinical issues. Recently, a number of biomaterials have been developed with improved thrombogenicity [2-5] but up to date, none of these biomaterials is totally thromboresistant.

The response of cell to surface topography has recently received increased attention and numerous studies on the effect of surface topography in several aspects of cell biology have been published [3,6-9]; particularly in areas of orientation [9], cell adhesion [10], morphology [11], cytoskeletal arrangement [12], proliferation rate and gene expression [6]. In contrast, the effect of substrate topography on platelet adhesion has not received as much attention and there is limited literature in the field. A recent study [13] showed that sub-micron polyurethane features can affect platelet adhesion when compared with pristine surfaces, but attempts were not made to optimized the topography to decrease adhesion. Carbon nanotube composite with topographically modified surfaces have also been demonstrated [14] with reduced platelet adhesion when compared to pristine surfaces.

In this work, platelet adhesion on a nano topographic polymeric surface is investigated. Nano fillers were fabricated by infiltrating a poly(D,L -lactide-*co*-glycolide) (PLGA) solution into a nano porous anodized alumina (NPAA) template. The effect of fillers' aspect ratio on platelet adhesion is investigated to determine the optimized geometrical parameters influencing the platelet adhesion.

EXPERIMENTAL
Nano Porous Anodized Alumina (NPAA) template
A monodisperse silica colloid with particle diameter ~270 nm was used to indent the surface of an aluminum foil (99.999%, Alfa Asear) for the fabrication of an organized NPAA [15]; using pressure at 4000 psi for 2 mins. The foil was then cleaned in an ultrasonic water bath to release the silica colloidal spheres attached to the surface. Anodization of the Al film was performed at 3 °C in 0.3 M H_3PO_4 aqueous solution, with voltage at 112 V.

Fabrication of textured PLGA films
Templates were cleaned in acetone to remove traces of contaminants and rinsed in deionised water before oven dried. The templates were then treated with 3,3,3-Trifluoropropyl-trimethoxysilane in the vapour-phase as anti-adhesion agent to facilitate the removal of infiltrated PLGA films. PLGA pellets were dissolved in dichloromethane (DCM) at the ratio of 1:25 w/v. The solution was then poured onto the prepared templates and vacuum baked at 55 °C for 5 days before releasing the nanotopographic features from the templates.

Platelet adhesion study on textured PLGA film
Fibrinogen from human plasma (Hfg) was preadsorbed onto all samples before incubation in platelet-rich-plasma (PRP). The prepared textured PLGA films were incubated in platelet suspension for 2 hrs at 37°C. The samples were then rinsed in phosphate buffered saline (PBS) to wash off the non-adhered platelets and fixed with 2.5% glutaraldehyde for at least 2 hrs at 4°C. Lastly, the samples were rinsed in PBS and through a series of graded alcohols before they were freeze dried at -80°C for Field Emission Scanning Electron Microscope FESEM observation (JEOL JSM-6340F) for determination on the amount adherent platelets. Experiments were performed in three independent experiments. The PRP and plasma from the repeated experiments were all from the same donor. The PRP and plasma are always stored in -20°C before use and is stable up to 12 months.

Quantification of fibrinogen adsorption by ELISA

The adsorption of fibrinogen on nanostructure substrates and pristine PLGA as control was quantified by the enzyme-linked immunosorbent assay (ELISA) array [16]. A calibration curve was established before quantification. The procedure is as follows: samples of 6-mm in diameter was first equilibrated in PBS for 2 hrs at room temperature in mini centrifuge tubes. Subsequently, the samples were placed in 96-well microplates with Hfg concentration of 0.05mg/ml for 1hr at 37°C. The samples were rinsed 5 times with buffer (PBS containing 0.1% (vol/vol) Tween 20) before transferring the samples to new wells on the microplate. The samples were then incubated in monoclonal antibody for 1 hr at 37°C. The samples were rinsed subsequently for 5 times before were transferred into new microplate wells. Finally, a colorimetric substrate of 3,3',5,5'-Tetramethylbenzidine was introduced into each well and left to incubate in the dark for 20 mins. Subsequently, 100μl of 0.5M H_2SO_4 was added to stop the enzymatic reaction. The solutions were quickly aspirated using a multiple pipette and transferred to new wells, where the absorbance values of the solutions was measured on a microplater reader at 450 nm within 5 mins. Adsorption experiments were done in triplicate (n= 3).

RESULTS & DISCUSSION

The dimensions of the topographic features created on the PLGA films from NPAA templates are shown in Table 1.

Table 1: Average topographic dimensions of the fabricated textured PLGA films.

Sample	Interspacing length-wise (nm)	Interspacing breadth-wise (nm)	Diameter (nm)	Height (nm)	Aspect Ratio
PLGA-1	129±21	147±28	94±21	210±56	2.2
PLGA-2	119±28	112±16	97±12	372±64	3.8
PLGA-3	125±17	118±28	87±22	427±92	4.9
PLGA-4	135±32	112±20	96±10	550±350	5.7

The number of platelets adhered on the substrates was quantified visually in FESEM images, where the total number of platelets observed (from average of 6 images) was divided by the surface area (n=3) expressed in mm^2. The amount of platelet adhesion found on the substrates in ascending order is as depicted in Figure 1: PLGA-4 < PLGA-3 < PLGA-2 < pristine PLGA < PLGA-1. The amount of fibrinogen adsorption on the PLGA-1 (most adherent platelets) and PLGA-4 (least adherent platelets) correlates respectively in accordance to the aspect ratio as shown in Figure 2.

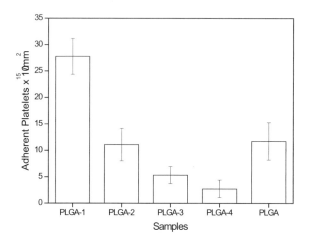

Figure 1: Graph displaying the amount of adherent platelets on the various PLGA films. Data are expressed as mean ± standard deviation of 3 independent experiments.

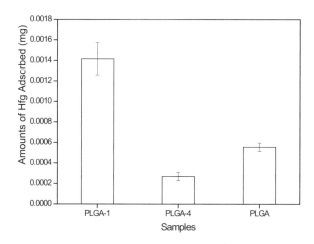

Figure 2: Graph displaying the highest and lowest amount of adsorbed fibrinogen relative to a pristine PLGA film for control. Data are expressed as mean ± standard deviation of 3 independent experiments.

Effect of aspect ratio on platelet adhesion and fibrinogen adsorption
The casted PLGA fillers from the NPAA template have relative constant interspacing and diameter, but the fillers' height was changed using different NPAA templates where the duration of anodization time was controlled accordingly. From the quantification experiments (figure 1) it can be seen that high levels of platelet adhesion were found on low aspect ratio of nano fillers (figure 3a), whereas the PLGA nano fillers with high aspect ratio (figure 3b) exhibited very low amount of platelet adhesion. Moderate platelet adhesion was observed on PLGA-2 (figure 3c & 3d) and PLGA-3 (figure 3e). We observed that fillers with high aspect ratio have the tendency to converge together after the incubation and drying processes (figure 3d). This occurrence could be induced by the freeze drying process under the influence of surface tension or heat during the FESEM observation. Nonetheless, we presume that these fillers exist individually separated in liquid phase during the PRP incubation. Being flexible, they could experience certain motion aiding on platelet-surface detachment, thus resulting in the low adhesion on these substrates. The low level of platelet adhesion on the high aspect ratio topography could be also due to the low level of adsorption or conformational changes of fibrinogen on these surfaces (figure 2, PLGA-4). Both effects, i.e. topography and fibrinogen content may act jointly to decrease the adhesion of platelets. In this study, we focused on mimicking the *in vivo* conditions where fibrinogen is always present in blood. More experiments are in plan to weigh the influence of each parameter. Comparatively, the pristine PLGA film displayed the highest platelet adhesion (figure 3f) when compared with the rest of the nanostructured PLGA films except for PLGA-1, signifying the importance of topographical control in the regulation of platelet response on substratum.

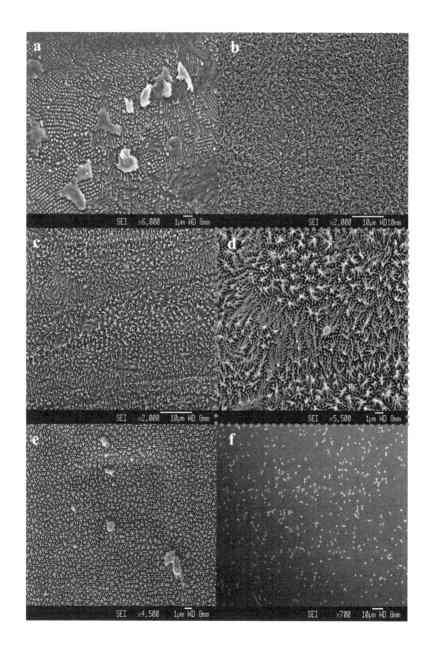

Figure 3: FESEM micrographs showing the effect of aspect ratio on platelet adhesion on: a) PLGA-1 with highest platelet adhesion when compared to pristine PLGA, b) PLGA-4 with lowest platelet adhesion, c) PLGA-2 with moderate platelet adhesion, d) magnified view of PLGA-2, e) PLGA-3 with moderate platelet adhesion and f) pristine PLGA showing highest platelet adhesion when compared with the rest of nanotopographic PLGA films except for PLGA-1.

CONCLUSIONS

We have shown evidence that a nano topographic modification on the surface of PLGA significantly lowers the level of platelets' adhesion. The lowest platelet adhesion as well as fibrinogen adsorption was observed on the nanotopographic surface with highest aspect ratio. The result demonstrated that controlling aspect ratio structures can be one important parameter influencing the amount of platelet adhesion which could be important in the design for low thrombogenic surfaces in blood contacting medical devices.

REFERENCES

1.   J. M. Grunkemeier, W. B. Tsai, M. R. Alexander, D. G. Castner, and T. A. Horbett, *J Biomed Mater Res* **51**, 669 (2000).

2.   J. M. Courtney, N. M. K. Lamba, S. Sundaram, and C. D. Forbes, *Biomaterials* **15**, 737 (1994).

3.   R. G. Flemming, C. J. Murphy, G. A. Abrams, S. L. Goodman, and P. F. Nealey, *Biomaterials* **20**, 573 (1999).

4.   S. L. Goodman, K. S. Tweden, and R. M. Albrecht, *J Biomed Mater Res* **32**, 249 (1996).

5.   D. F. Gibbons, *Annu Rev Biophys Bioeng* **4**, 367 (1975).

6.   R. Singhvi, G. Stephanopoulos, and D. I. C. Wang, *Biotechnol Bioeng* **43**, 764 (1994).

7.   A. Curtis, presented at the Symposium on Nanoscale Materials Science in Biology and Medicine held at the 2004 MRS Fall Meeting, Boston, MA, 2004 (unpublished).

8.   Yoshihiro Ito, *Biomaterials* **20**, 2333 (1999).

9.   A. Curtis and C. Wilkinson, *Biomaterials* **18**, 1573 (1997).

10.   Sharma Chandra P., Chandy Thomas, and M. C. Sunny, *J Biomater Appl* **1**, 533 (1986).

11. S. Heydarkhan-Hagvall, C. H. Chof, J. Dunn, S. Heydarkhan, K. Schenke-Layland, W. R. Maclellan, and R. E. Beygul, *Cell Commun Adhes* **14**, 181 (2007).

12. M. J. Dalby, M. O. Riehle, D. S. Sutherland, H. Agheli, and A. S. G. Curtis, *Eur J Cell Biol* **83**, 159 (2004).

13. K. R. Milner, A. J. Snyder, and C. A. Siedlecki, *J Biomed Mater Res A* **76A**, 561 (2006).

14. L. B. Koh, I. Rodriguez, and J. Zhou, *J Biomed Mater Res A* **86A**, 394 (2008).

15. S. Fournier-Bidoz, V. Kitaev, D. Routkevitch, I. Manners, and G. A. Ozin, *Adv. Mater.* **16**, 2193 (2004).

16. A. Higuchi, K. Sugiyama, B. O. Yoon, M. Sakurai, M. Hara, M. Sumita, S. Sugawara, and T. Shirai, *Biomaterials* **24**, 3235 (2003).

Mater. Res. Soc. Symp. Proc. Vol. 1140 © 2009 Materials Research Society   1140-HH05-05

Electro-Wetting and ArF Excimer Laser Induced Photochemical Surface Modification of Hydrophilic and Hydrophobic Micro-Domain Structure on IOL Surface for Blocking After Cataract

Y. Sato[1], K. Kawai[2], M. Sasoh[3], H. Ozaki[4], T. Ohki[5], H. Shiota[5] and M. Murahara[1]

[1]Entropia Laser Initiative, Tokyo Institute of Technology
[2]School of Medicine, Tokai University
[3]School of Medicine, University of Mie
[4]School of Medicine, Fukuoka University
[5]School of Medicine, University of Tokushima

ABSTRACT

A hydrophilic and hydrophobic micro-domain structure was formed photo-chemically on intraocular lens [IOL] surface by, patterned ArF excimer laser irradiation.   For this new method, the new IOL was developed to inhabit a fibrin attachment.

The water was poured into the thin gap between the silica glass and the IOL with capillary phenomenon and applying altenated current [A.C] voltage between IOL and fused silica glass. In this condition, the ArF excimer laser light is projected through a patterned mask in reduced size. In order to evaluate the wettability of the modified sample, the contact angle with water was measured.   The contact angle with water decreased from 110 degrees for the untreated silicone IOL to 72 degrees at the laser fluence of 30 mJ/cm^2 and shot number of 15000.   However, at the laser fluence 10 mJ/cm^2 and shot number of 12000 applying the A.C. voltage of 1000 V,   the water contact angle became to 43 degrees.

INTRODUCTION

Plastics are clinical applied for ophthalmic implant devices as contact lens [CL], IOL and artificial cornea.   Silicone rubber, poly-hydroxyethyl methacrylate [HEMA] and polymethylmethacrylate [PMMA] lens has been used for an intraocular lens because of its high transmittance in the visible region, inert material and biocompatibility.   However, when these materials were implanted in the living body for long term, a protein called fibrin is adsorbed onto the material surface to proliferate epithelial cells and cause clouding [1].   Therefore, there are many previous reports on IOL surface modified into hydrophilic or hydrophobic property by various methods.

David developed the super hydrophobic surface with 100-micrometer roughness on silicon surface by lithography technique [2].   Satriano and Marletta incorporated oxygen groups to be hydrophilic on a poly(hydroxymethylsiloxane) [PHMS] surface by O_2-plasma and Ar^+-ion beam irradiations and cultivated fibroblast on the modified surface [3].   In order to inhibit cell adhesion, Okano and coworkers formed a micro-phase separated structure of hydrophilic and hydrophobic domains by polymerization and investigated blood protein adsorption.   The sample surface with a micro-phase separated structure of hydrophilic and hydrophobic domains to inhibit protein adsorption [4].   These reports indicate that hydrophilic surface is desirable to improve biocompatibility, and a hydrophobic one is required to hinder protein adsorption. Thus, we demonstrated the hydrophilic and hydrophobic micro domain structure on silicone IOL surface by the ArF excimer laser.

In our previous study, we incorporated desired functional groups or metal atoms onto the polymer surface using an excimer laser to generate a hydrophilic or hydrophobic surface and found that especially for PTFE, B, Al and H atoms were effective in defluorinating [6,7]. Having studied the compatibility between the defluorinating atoms and a living body extensively, it was found that when using B and Al, their compounds remained on the modified surface and became impurities to cause rejections;   however, no rejection occurred on the hydrophilic PTFE modified with H, i.e., water (H_2O) [8].   This was confirmed by post-operated days [POD] test on the sample that had been implanted in the cornea of a rabbit.   Moreover, hydrophilic and hydrophobic micro domain structure was formed on the PMMA IOL surface by the ArF excimer laser irradiation. As the results, micro domain structure inhibited the protein adsorption [9]. Based on these studies, an electro-wetting method is used in combination with forming the micro domain structure of hydrophilic and hydrophobic on silicone IOL to improve the efficiency of photochemical surface modification

EXPERIMENTAL

Figure1 shows the experimental procedure for photo-chemical surface modification with an electro-wetting method.   Water is placed on the IOL and the fused silica glass is placed on top of it to make a thin water layer by capillary phenomenon and applying the A.C.voltage between the IOL and fused silica glass.   Then the A.C. voltage was applied between the IOL and the fused silica glass.   In this condition, the ArF excimer laser light is projected through a patterned mask in reduced size.   The water is photo-dissociated by the laser photon to produce -H and -OH radicals, as shown in Equation 1. at the same time, the C-H bond of silicone IOL is also photo-excited because the photon energy (147 kcal/mol) of ArF excimer laser is higher than the binding energy of C-H (81 kcal/mol).   The H radicals pulled out the H atoms of silicone IOL surface to produce H_2 gas, as shown in Equation 2.   As the results, the H atoms are replaced with –OH functional groups, which modified the silicone IOL surface into hydrophilic.

$$H_2O + h\nu \rightarrow H^* + OH^* \quad (\text{h } \nu < 242nm) \qquad (1)$$

$$-(-O-Si(CH_3)_2-O-)n + H^* + OH^* + h\nu \rightarrow (-O-Si(CH_2-OH)_2-O-)n + nH_2 \qquad (2)$$

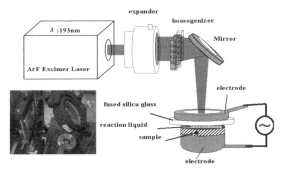

Fig.1 Experimental procedure

RESULTS AND DISCUSSIONS

The contact angle with water was measured to evaluate wettability of applying the AC voltage. Figure 2 shows the A.C. voltage dependence upon the water contact angle on the silicone IOL . The contact angle of the silicone IOL was 110 degrees. The contact angle of water decreased from 110 degrees for the untreated sample to 35 degrees by applying the A.C. voltage [11-12].   At 1000 V, the water contact angle became minimum to 35 degrees on the silicone IOL.

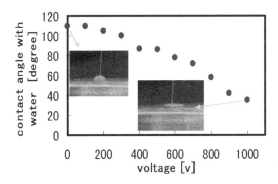

Fig.2 Applying the AC voltage dependence on the water contact angle on silicone IOL surface

Figure 3 shows the correlation between water contact angle and laser shot number with and without applying the A.C.1000 V. Of the sample modified with laser irradiation only, the contact angle became 70 degrees from 110 degrees for untreated samples at the laser fluence of 30 mJ/cm^2 and shot number of 15000. And a threshold of hydrophilic group incorporation was 9000 shots. By the combine A.C. voltage and the ArF excimer laser treatments, however, the threshold of hydrophilic group incorporation became small. It is considered that the water layer was thinned by the electro-wetting. As the result, the excited efficiency on the IOL surface became higher. And the water contact angle also became minimum to 42 degrees at 10 mJ/cm^2 and shot number of 12000.

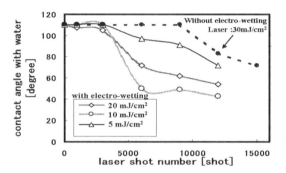

Fig.3 Correlation between the laser shot number and contact angle with water

The incorporation of functional groups on the silicone IOL was evaluated by attenuated total reflectance Fourier-transform infrared spectroscopy [ATR FT-IR] using a germanium crystal before and after laser irradiation. Figure 4 shows the IR spectra of untreated and sample modified into hydrophilic. The spectrum of the untreated sample indicated the absorption of Si-O stretch in the region of 1100 cm^{-1}. On the other hand, the spectrum of the treated sample presented the absorption of OH group in the region of 3300cm^{-1}. It was clarified by the ATR-FT-IR analysis that the OH functional groups were incorporated on the silicone IOL surface.

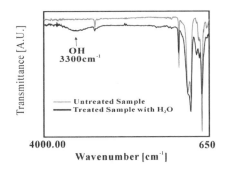

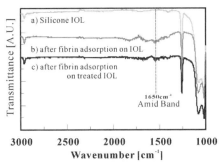

Fig.4　ATR-FT-IR analysis for treated IOL　　　Fig.5 Fibrin adsorption test by ATR-FT-IR

The hydrophilic group was incorporated at interval of 20 micrometer to form the micro domain structure on the silicone IOL surface [9].　The contract angle reached to 75 degrees from 110 degrees with electro-wetting of 1000 V at the laser fluence of 10 mJ/cm^2 and shot number of 12000.　These samples were soaked in the 0.1 % fibrin solution for 48 hours at 36 degree Celsius.　After rinse with waters, the absorption coefficient of amide band for fibrin at 1650 cm^{-1} was measured on the sample surface by ATR-FT-IR, as shown table 1 and figure 5. (a) ,(b) and (c) means the silicone IOL, after fibrin adsorption on untreated IOL and fibrin adsorption on treated IOL.　As the results, the absorption coefficient of the amide band on the modified surface became 0.0004, decreasing to three-five of the non-treatment sample's.

Table.1 fibrin adsorption test on treated and untreated IOL surface

	Laser [shot]	Contact angle [degree]	Abs. Coefficient
Silicone IOL	0	110	0.001
Treated sample	6000	88	0.0003
Treated sample	12000	74	0.0004

Conclusion

The hydrophilic groups were photo-chemically incorporated on the silicone IOL surface by the combined electro-wetting and the ArF excimer laser.　As the results, it was found that the photochemical reactions in surface modification process were activated to 20 times by the electro-wetting.　At the 5 mJ/cm2 and 12000 shot, the water contact angle on the silicone IOL became 42 degrees from 110 degrees.　Furthermore, the hydrophilic and hydrophobic micro-domain structure was inhibited the fibrin sticking with electro-wetting and ArF excimer laser irradiation

Acknowledgement

This research was supported in part by grant #19659446 from the Ministry of Education and science JAPAN.

Reference

1.　Janice L. Bohnerr,* Thomas A. Horbetr,* Buddy D. Rarner,* and Frederick H. Roycef, Investigative *Ophthalmology & Visual Science*, **29** .362-373 (1988)

2.　David Quere, *Nature Material*, **1** 14-15 (2002)

3.　C. Satriano and G. Marletta, *J. Mater. Sci., Mater. in Medicine* **14**, 663-670 (2003)

4.　N. Ogata, K. Sanui, H. Tanaka, Y. Takahashi, Y. Sakurai, T. Akaike, T. Okano and K. Kataoka, *J. Appli. Poym. Sci.*, **26**, 2293-2303, 1981

5.　P.K. Chua,, J.Y. Chen, L.P. Wang, N. Huang, Materials Science and Engineering R 36 143–206 (2002)

6.　Y. Sato and M. Murahara *J. Adhesion Sci. and Technol*, **18**, 1687-1697 (2004)

7.　M. Okoshi and M.Murahara, *App. Phys. Lett.* **72**, 2616-2618 (1998)

8.　Y. Sato, J. M. Parel and M. Murahara, *Mater. Res. Soc. Symp. Proc.* **711**, 277-282 (2002)

9.　Y. Sato and M. Murahara, *J. Adhesion Sci. and Technol.*, **18**, 1545-1555 (2004)

10.　Robert A.H and B.J. Feensta, *Nature*, **425** 25 (2003)

11.　H.Anai, Y.Sato and M.Murahara, *Mat. Res. Soc. Symp. Proc.*, **890,** 229-234 (2001)

Synthesis and Characterization of Nanomaterials for Biomedical Applications

Mater. Res. Soc. Symp. Proc. Vol. 1140 © 2009 Materials Research Society   1140-HH05-07

Controlled Actuation of Shape-Memory Nanocomposites by Application of an Alternating Magnetic Field

M. Y. Razzaq,[1] M. Behl,[1] K Kratz,[1,2] A. Lendlein[1,2]

[1]Center for Biomaterial Development, Institute of Polymer Research, GKSS Research Center Geesthacht, Kantstr. 55, 14513 Teltow, Germany
[2]Berlin-Brandenberg Center for Regenerative Therapies Augustenburger Platz 1, 13353 Berlin, Germany

ABSTRACT

Shape-memory polymers are smart materials with an high application potential as intelligent implant material e.g. as self adjusting orthodontic wires or selectively pliable tools for small scale surgical procedures. Typically heat or light are used to initiate the shape-memory effect. By incorporating magnetic nanomaterials into shape-memory polymers a non-contact triggering of shape-memory polymers can be realized when the nanocomposite is exposed to an alternating magnetic field.

Here we report controlled actuation of shape-memory polymer networks by application of an alternating magnetic field. Shape-memory nanocomposites were prepared by crosslinking of oligo(ε-caprolactone)dimethacrylate in the presence of silica coated nanosized magnetite nanomaterials. Effects of magnetite nanomaterials on the thermal and mechanical properties of the composite have been investigated by means of differential scanning calorimetry (DSC), dynamic mechanical analysis at varied temperature (DMTA) as well as by tensile tests and cyclic thermomechanical experiments. The influence of the nanomaterials on the shape-memory and elastic properties of the nanocomposite are discussed. Finally, the macroscopic shape-memory effect by inductive heating of the nanocomposite samples in an alternating magnetic field ($f = 258$ kHz; $H = 17$ kA/m) was explored.

INTRODUCTION

Shape-memory materials that can be triggered by a magnetic field have a high application potential in medical devices. The material could lead to medical implants such as polymer stents, which could be inserted in a compact temporary shape, and then remotely triggered to achieve their final shape. Another important area of application could be on demand drug delivery from an implanted depot which could be triggered non-invasively by a magnetic field. [1] The main advantage of this method of remote actuation over conventional heating is the fact that the magnetic nanoparticles can be heated in a very controlled manner by using a magnetic field of suitable field strength and frequency. [2, 3] Here, we report about the magnetic actuation of a shape-memory polymer (SMP) network with crystallizable switching domains loaded with magnetic nanomaterials. A crosslinked polymer network was selected to reach high shape recovery rate and avoid relaxation effects. The polymer networks were prepared from oligo(ε-caprolactone)dimethacrylate (PCLDMA), silica coated magnetite nanomaterials (mag-NM) and a thermosensitive initiator. In such a polymer network the crystalline domains formed by the PCL chain segments are used to temporarily fix a mechanical deformation. T_m related to these domains ($T_{m,PCL}$) will substantially influence the switching temperature (T_{sw}) of the shape-memory effect. [4] We have investigated the effects of different loadings of mag-NM on the thermal, mechanical and shape-memory properties of the nanocomposites (NC-PCL) along with the controlled non-contact triggering of the NC-PCL by exposure to an alternating magnetic field.

EXPERIMENTAL

PCLDMA was prepared by equimolar conversion of terminal hydroxyl groups of the oligo(ε-caprolactone)diol with M_n = 8300 g·mol^{-1}, M_w/M_n = 1.4 (GPC) (Solvay Chemicals, Warrington, UK) with 2-isocyanatoethyl methacrylate (Sigma-Aldrich) in 88 % yield. A degree of methacrylation (D_a) of 93% was determined by ^1H NMR spectroscopy, resulting in 86% dimethacrylated, 13% monomethacrylated and 0.5% non-methacrylated oligo(ε-caprolactone)diol. [5] Nanomaterials consisting of an iron(III)oxide core in a silica matrix (AdNano MagSilica 50) were provided by Degussa Advanced Nanomaterials (Degussa, Hanau, Germany) as a powder. The mean aggregate size was 90 nm, the mean domain size was 20-26 nm, and the domain content was 50-60 wt%. [6] Nanocomposites containing 5 and 10 wt% of mag-NM (NC-PCL05, NC-PCL10) were prepared by heating a mixture of PCLDMA and mag-NM at 80 °C under vigorous stirring at 200 rpm (Stirrer: Heidolph RZR 2102) for ten minutes. After addition of 1.5 mol% benzoyl peroxide the highly viscous mixture was transferred between two glass plates (46×46 mm^2) which were separated by a Teflon^{TM} spacer with a thickness of 1 mm and was cured at 80 °C for 24 h. To determine the gel content (G) and the degree of swelling (Q), the film was swollen in a 100-fold excess of chloroform overnight and dried at room temperature. This procedure was repeated until constant weight was obtained. [4]

Scanning transmission electron microscopy (STEM) was performed using scanning electron microscope (SEM) Zeiss Gemini, Supra 40VP (Zeiss, Jena Germany) at an accelerating voltage of 10 kV. Ultra thin sections of 70 nm were cut on a Leica FC6 cryo ultra-microtome at 130 °C (Leica Microsystems GmbH Wetzlar, Germany). Investigation by STEM was carried out by a STEM extension installed onto the SEM-specimen stage. Wide angle X-ray scattering (WAXS) measurements were performed on a X-ray diffraction system Bruker D8 Discover with a two-dimensional detector from Bruker AXS (Karlsruhe, Germany). DSC experiments were performed with a Netzsch (Selb, Germany) DSC 204, in the temperature range from -100 to 100 °C with a constant heating and cooling rate of 10 K·min^{-1}.

The determination of the dynamic mechanical properties was performed on a DMTA Eplexor 25N (Gabo, Ahlden, Germany) in temperature-sweep mode from -100 to 100 °C at a constant heating rate of 2 K·min^{-1} with an oscillation frequency of 10 Hz. Tensile tests at room temperature were carried out on Zwick (Ulm, Germany) Z005 and Z2.5 tensile testers. Experiments at elevated temperature and the cyclic thermo-mechanical tests were performed on Zwick Z1.0 and Z005 tensile testers equipped with thermo chambers and temperature controllers (Eurotherm Regler, Limburg, Germany). In each experiment the strain rate was 5 mm·min^{-1}. Shape-memory properties were determined in a stress-controlled cyclic thermo-mechanical test by elongating the sample at T_{high} = 80 °C to ε_l with subsequent cooling to T_{low} = 0 °C while the stress was kept constant. Then stress was lowered to zero resulting in the temporary shape of the sample at strain ε_u. Finally, the shape-memory effect was observed by reheating to T_{high} with a heating rate of 2 K·min^{-1}. Each experiment consisted of three cycles. From each cycle the shape fixity rate (R_f) and shape recovery rate (R_r) were determined. [7] Similarly the bending recovery rate (R_b) was determined by bending the samples to an angle of 90° at T_{high} with a subsequent cooling to T_{low}. Shape recovery was observed by heating the deformed sample to T_{high} by exposing to magnetic field or by increasing the environmental temperature (T_{env}) and the change of the angle was recorded. Finally the bending recovery rate (R_b) was calculated from the ratio of different angles before and after recovery. [8]

Inductive-heating of silica coated iron(III) oxide particles was accomplished by positioning the sample in an alternating magnetic field at a frequency of $f = 258$ kHz. By adjusting the generator (TIG 5/300 high-frequency generator, Huettinger Electronic; Freiburg, Germany) power output, the magnetic field strength in the center of the coil could be varied between 7 and 30 kA/m.

RESULTS AND DISCUSSIONS

Morphology of the nanocomposites

The value of G for all the samples was between 90 and 96%. The degree of swelling (Q) increased by increasing mag-NM content, indicating a decrease in crosslink density induced by the nanoparticles (Table 1).

The nanocomposite distribution was investigated by means of SEM and was found to be macroscopically homogenous. The mag-NM concentration was uniform at the top and bottom of the sample. No sedimentation of the nanomaterial was observed. In swelling experiments no influence of the composite distribution could be determined in the SEM measurements. However, STEM (Figure 1a) results display certain inhomogenities at microscopic level. Agglomerates of varying size similar to those found in the bulk mag-NM powder were detected in the nanocomposites.

Figure 1b shows the WAXS patterns of homonetworks (N-PCL00) and nanocomposites as a function of contents of mag-NM. Two distinct diffraction peaks are observable for N-PCL at $2\theta = 21.5°$ and $2\theta = 23.8°$, which were attributed to be (110) and (200) planes, of an orthorhombic crystalline structure of PCL. [9] In the nanocomposites no change of the peak positions could be determined. However, it can be observed that the scattering intensity of the diffraction peaks steadily decreases with an increase in the loading level of mag-NM. The degree of crystallinity

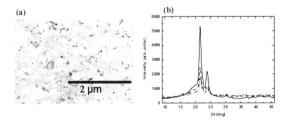

Figure 1: (a) STEM image of NC-PCL05 containing 5 wt% of mag-NM. (b) WAXS patterns of N-PCL and nanocomposites. (solid lines: N-PCL00, dash lines: NC-PCL05, dash-dot lines: NC-PCL10)

Table 1. Thermal and morphological properties of N-PCL00 and nanocomposites.

Sample ID	G [%] [a]	Q [%] [b]	T_m [°C] [c]	$T_{tan\,\delta,\,max}$ [°C] [d]	ΔH_m [J·g^{-1}] [e]	DOC [%] [f]
N-PCL00	95.3 ± 0.7	643 ± 1	52 ± 1	- 42 ± 2	53.6 ± 0.7	49.1 ± 0.6
NC-PCL05	94.6 ± 0.1	676 ± 9	51 ± 1	- 41 ± 2	49.5 ± 2.9	36.4 ± 0.4
NC-PCL10	95.3 ± 0.2	700 ± 2	52 ± 1	- 40 ± 2	48.8 ± 3.7	32.1 ± 0.3

[a] G is gel content [b] Q is the degree of swelling [c] T_m is melting temperature determined from melting peak of DSC. [d] $T_{tan\delta,\,max}$ is the glass transition temperature determined by DMTA from the tan delta peak. [e] ΔH_m is the melting enthalpy (corrected for particle loading) determined by DSC. [f] Degree of crystallinity determined by WAXS measurements.

was found to decrease consistently by increasing the mag-NM content, indicating a disturbing effect of mag-NM in the crystallization of the PCL phase. (Table 1)

Thermal and thermo-mechanical properties

The thermograms obtained from DSC indicate that the $T_{m,PCL}$ at 52 °C is not significantly affected by the incorporation of mag-NM (Table 1. Figure 2a). However, the melting enthalpy (ΔH_m) of the nanocomposites decreased with increased mag-NM content because of the decreased degree of crystallinity .

The dynamic mechanical analysis at varied temperature represents a decrease in storage modulus *(E')* with temperature for all the samples. (Figure 2b) The steep decline of the curves around the $T_{m,PCL}$ represents the temperature range over which the samples are softening. A slight increase in the storage modulus was observed from -50 to 60 °C by increasing the percentage of mag-NM, indicating the increasing stiffness of the nanocomposites. A sharp *tan δ* peak related to the T_g of the amorphous PCL was observed at around -42 °C for all samples. Inclusion of mag-NM causes a decrease of the tan *δ* peak. Additionally a slight increase in the peak temperature of glass transition relaxation was observed at increased mag-NM content indicating the increasing rigidity of the nanocomposites (Table 2).

Tensile tests for the N-PCL00 and the nanocomposites have shown an increase in elastic moduli *(E)* of the resulting composites with increasing mag-NM content at 25 °C. However, at 80 °C the value of *E* decreases when the particle content is increased. At 80 °C, the material is significantly softer as the PCL domains are completely amorphous. The elongation at break (ε_R)

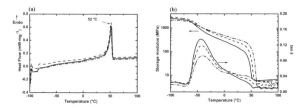

Figure 2: (a) DSC spectra (b) Dynamic mechanical analysis at varied temperature (solid lines: N-PCL00, dash lines: NC-PCL05, dash-dot lines: NC-PCL10)

Table 2. Mechanical properties of N-PCL00 and nanocomposites determined by tensile tests and DMTA.

Sample ID	25 °C				80 °C			
	E [MPa] [a]	σ_m [MPa] [b]	ε_R [%] [c]	E' [MPa] [d]	E [MPa] [a]	σ_m [MPa] [b]	ε_R [%] [c]	E' [MPa] [d]
N-PCL00	136 ± 8	10.4 ± 0.3	222 ± 45	150 ± 2	9.6 ± 2.1	2.7 ± 0.1	34 ± 1	3 ± 2
NC-PCL05	169 ± 4	11.5 ± 0.3	148 ± 28	251 ± 2	5.6 ± 0.6	1.6 ± 0.2	74 ± 9	5 ± 2
NC-PCL10	284 ± 21	12.8 ± 2.3	9 ± 2.3	332 ± 2	7.0 ± 2.4	1.4 ± 0.1	71 ± 2	6 ± 2

[a] E is the elastic modulus, [b] σ_m is maximum stress, [c] ε_R is elongation at break, [d] E' is storage modulus determined by DMTA.

is decreased for samples having higher nanomaterial contents. At 80 °C, ε_R increases with increased mag-NM content. (Table 2) Compared with the storage moduli obtained by DMTA, Young's moduli determined by tensile tests were lower. This might be because of the different measurement principles or might be typical for these particles. [2]

Shape-memory properties

Magnetically-induced shape-memory effect is demonstrated in figure 3 for nanocomposite sample NC-PCL10, where a change of the temporary (bend bar) to the permanent (straight bar) shape within 22 s is shown. Additionally, the shape-memory properties were quantified in cyclic thermomechanical tests. The addition of mag-NM results in lower strain recovery ratio (R_r) but slightly increased value of strain fixity ratio (R_f) (Table 3). R_r is controlled by the elastic recovery force, which depends on the length of the chain segment and the mobility of the chains. By increasing the mag-NM content the movements of the amorphous chain segments are more and more restricted causing a decrease in R_r. The increase of the mag-NM content causes a slight increase in the R_f, as the mobility of the amorphous chain segments is restricted, while the effect of reduced crystallization with increased nanomaterial content becomes less significant. In contrast to R_r, the high values of bending recovery ratio (R_b) were obtained for thermally- as well as magnetically-induced recoveries and no significant change in R_b was observed by increasing the mag-NM contents.

Table 3. Shape-memory properties determined in cyclic, thermomechanical tensile tests and bending tests.

Sample ID	Thermally-induced recovery					Magnetically-induced recovery
	$R_{f(1)}$ [%] [a]	$\overline{R}_{f(2-3)}$ [%] [b]	$R_{r(1)}$ [%] [c]	$\overline{R}_{r(2-3)}$ [%] [d]	$\overline{R}_{b(2-3)}$ [%] [e]	$\overline{R}_{b(2-3),\,mag}$ [%] [f]
N-PCL00	93.0	93.1 ± 0.7	94.6	99.8 ± 0.1	99.9 ± 0.1	-
NC-PCL05	96.9	97.2 ± 0.1	91.6	99.6 ± 0.3	99.8 ± 0.2	100
NC-PCL10	98.3	98.3 ± 0.2	75.0	95.6 ± 2.1	99.9 ± 0.1	100

[a] $R_{f(1)}$ is the strain fixity ratio of the first cycle, [b] $\overline{R}_{f(2-3)}$ is the average strain fixity ratio of the second and third cycle, [c] $R_{r(1)}$ strain recovery ratio of the first cycle, [d] $\overline{R}_{r(2-3)}$ is the average strain recovery ratio of the second and third cycle. [e] $\overline{R}_{b(2-3)}$ is the average bending recovery ratio for the second and third bending cycle. [f] $\overline{R}_{b(2-3),mag}$ is the average bending recovery ratio for the second and third cycle by inductive-heating in the magnetic field.

Figure 3: Series of photographs showing the macroscopic shape-memory effect of NC-PCL10 with 10 wt% mag-NP. The picture series shows a transition from temporary to permanent shape in a magnetic field of $H = 17$ kA/m and $f = 258$ kHz.

CONCLUSIONS

Magnetically-induced non-contact triggering of the shape-memory effect has been demonstrated for nanocomposites from PCLDMA and silica coated mag-NM. Thermal properties of the polymer networks were shown to be independent of the nanomaterial content, but the degree of crystallization was reduced significantly by the dispersed nanomaterials. The elongation at break ε_R was decreased with increased nanomaterial content at low temperature but at elevated temperature an opposite effect was observed. Cyclic thermomechanical tests have shown an increased R_f value from 93 % for homonetworks to 98% for the nanocomposite with 10 wt% nanomaterials. In contrast, R_r was slightly decreased to 95% with increased nanomaterial content. Magnetically- and thermally-induced bending recovery ratios were high and were quite comparable to each other.

ACKNOWLEDGMENTS

M. Behl acknowledges financial support by Helmholtz Association for a special support grant (SO-NG-054) for the cryo ultramicrotome.

REFERENCES

[1] E.R. Edelman, L. Brown, J. Kost, J. Taylor, R. Langer, Transaction American Society For Artificial Internal Organs **30**, 445 (1984).
[2] R. Mohr, K. Kratz, T. Weigel, M. Lucka-Gabor, M. Moneke, A. Lendlein, Proceedings of the National Academy of Sciences of the United States of America **103**, 3540 (2006).
[3] M. Y. Razzaq, M. Anhalt, L. Frormann, B. Weidenfeller, Materials Science and Engineering A-Structural Materials Properties Microstructure and Processing **444**, 227 (2007).
[4] A. Lendlein, A. M. Schmidt, M. Schroeter, R. Langer, Journal of Polymer Science Part a-Polymer Chemistry **43**, 1369 (2005).
[5] A. S. Karikari, W. F. Edwards, J. B. Mecham, T. E. Long, Biomacromolecules **6**, 2866 (2005).
[6] H. Gottfried, C. Janzen, M. Pridoehl, P. Roth, B. Trageser, G. Zimmermann, inventor. U.S. Patent. 6, **746**, 767 (2003).
[7] N. Y. Choi, S. Kelch, A. Lendlein, Advanced Engineering Materials **8**, 439 (2006).
[8] A. Lendlein, S. Kelch, Angewandte Chemie-International Edition **41**, 2034 (2002).
[9] A. Baji, S. C. Wong, T. X. Liu, T. C. Li, T. S. Srivatsan, Journal Of Biomedical Materials Research Part B-Applied Biomaterials **81B**, 343 (2007).

Mater. Res. Soc. Symp. Proc. Vol. 1140 © 2009 Materials Research Society   1140-HH03-02

Magnetron-Sputtered Zinc-Doped Hydroxyapatite Thin Films for Orthopedic Applications

Patrick C. Marti[1], Zoe H. Barber[1], Roger A. Brooks[2], Neil Rushton[2], Serena M. Best[1]

[1]Department of Materials Science & Metallurgy, University of Cambridge, Pembroke Street, Cambridge, CB2 3QZ, UK

[2]Orthopaedic Research Unit, University of Cambridge, Box 80, Addenbrooke's Hospital, Hills Road, Cambridge CB2 2QQ, UK

ABSTRACT

Magnetron co-sputtering was used to produce zinc-doped hydroxyapatite thin films on Ti-6Al-4V substrates containing 3.4 wt. % and 6.4 wt. % zinc, as measured by energy dispersive x-ray spectroscopy (EDS). X-ray diffraction (XRD) confirmed the formation of a crystalline apatite structure following heat treatment for 3 hours at 600°C. Scanning electron microscopy (SEM) showed no difference in the coating morphology upon the addition of zinc to the coating. Human osteoblast (HOB) cells were used to assess the biocompatibility of hydroxyapatite and zinc-doped hydroxyapatite films over 14 days. HOBs attached and grew well on all surfaces, with 3.4 wt. % zinc-doped hydroxyapatite films showing significantly enhanced cell proliferation. This result confirms that magnetron-sputtered zinc-doped hydroxyapatite thin films have significant potential as coatings on implants for orthopedic applications.

INTRODUCTION

Hydroxyapatite has attracted a great deal of interest for use in orthopedic biomaterial applications due to its favorable biological properties. However, the poor mechanical properties of bulk hydroxyapatite generally render it unsuitable for major load-bearing applications [1]. To address this problem, methods for coating metal implants with hydroxyapatite have been developed [2]. Hydroxyapatite-coated implants have been shown to improve bone growth and apposition *in vivo*, thus increasing the clinical success of the implant [3].

There are a number of different methods of producing hydroxyapatite coatings, with plasma spraying being the most common commercially. While plasma-sprayed coatings have enjoyed success clinically, studies have shown that plasma-sprayed coatings are prone to loss of mechanical integrity and can be resorbed within the body [4,5]. Magnetron sputtering provides a means for the production of dense, uniform thin hydroxyapatite coatings (<3 μm). Studies have shown that magnetron-sputtered hydroxyapatite thin films provide a favorable surface for cell attachment and growth, and improve the apposition of bone to an implant *in vivo* [6,7].

Recently, a number of studies by Thian *et al.* have documented the production of silicon-containing hydroxyapatite thin films through a magnetron co-sputtering process [8]. The work showed that the introduction of controllable levels of silicon into the film could enhance the attachment and growth of HOB cells on the film's surface. Zinc is a trace element found in bone that has been shown to have a stimulatory effect on osteoblast growth and an inhibitory effect on osteoclast resorption [9]. The biological properties of zinc-substituted calcium phosphates have been investigated, confirming the positive influence of zinc on the behavior of cells [10]. However, relatively few studies have examined the properties of zinc-containing hydroxyapatite coatings. The goal of this study was thus to investigate the physical, chemical, and biological properties of a range of magnetron co-sputtered zinc-doped hydroxyapatite thin films.

MATERIALS AND METHODS

Magnetron Co-Sputtering

Ti-6Al-4V discs (10 mm diameter) were ground with silicon carbide paper (to 1200 grit) before being ultrasonically washed in acetone, rinsed with distilled water, and dried with compressed air. Targets of phase-pure sintered HA and zinc were attached to water-cooled magnetrons 105 mm apart. The target to substrate distance was 44 mm. Sputtering runs were carried out for 4 hours at a pressure of 0.8 Pa. Radio-frequency power was supplied to the HA target at 60 W, and a direct current was applied to the zinc target with a power of either 1.5 W or 3.0 W. Samples were heat-treated following deposition at 600°C for 3 hours in an Ar/H_2O atmosphere to ensure crystallization of the coatings.

Physical and Chemical Characterization

Coating thickness was determined by masking a portion of a substrate and measuring the step created after removal of the mask with a Stylus Profilometer. A Camscan MX2600 operating at 10kV was used for SEM imaging and EDS analysis of each coating. Each sample was coated with a thin layer of carbon prior to analysis. A Philips PW 1730 X-ray generator with a voltage of 40 kV and current of 40 mA was used to generate Cu Kα radiation for XRD analysis. A divergence slit of 1° and receiving slit of 0.2 mm were used to scan each coating from 20-45°2θ with a step size of 0.025°2θ and dwell time of 20 s.

In Vitro Analysis

Human osteoblast (HOB) cells were cultured in McCoy's 5A medium containing 10% FCS, 1% PSG, and Vitamin C (30 μg/ml). Prior to cell culture experiments, heat-treated samples were sonicated for 30 minutes in ethanol and baked at 200°C for 2 hours to ensure sterilization. A 100 μl cell suspension was dropped onto each sample surface at a concentration of 10^4 cells/disc and incubated for 3 hours to allow for attachment prior to immersion in culture medium. The AlamarBlue™ assay was used to measure cell metabolic activity on each sample at days 1, 3, 7, and 14. At each time point, regular culture medium was replaced with medium containing 10% AlamarBlue™. After 4 hours of incubation, triplicate 160 μl samples of medium were pipetted into a black 96-well plate. The fluorescence of the samples was measured using a plate reader at a wavelength of 570 nm and a reference wavelength of 600 nm. Medium containing 10% AlamarBlue™ was used as a standard. The CyQUANT™ cell assay was used to measure cell number on each sample at days 1 and 14. At each time point, the cells were frozen, thawed, and suspended in cell lysis buffer prior to being incubated in GR dye for 5 minutes. Triplicate 100 μL samples of cell lysis/dye solution were read with a plate reader at a wavelength of 480 nm and a reference wavelength of 520 nm. The two-tailed t-test was used to determine the statistical significance of each set of quantitative data. The difference in results was said to be statistically significant for values of $p \leq 0.05$.

Cell organization and morphology were assessed using fluorescent microscopy at days 3 and 14. For fluorescent microscopy, cells were fixed in 4% paraformaldehyde and permeabilized prior to being stained for 1 hour with phalloidin-TRITC. Cells were mounted on glass slides using VECTASHIELD® mounting medium containing DAPI for nuclear staining.

RESULTS AND DISCUSSION

Physical and Chemical Characterization

The results of semi-quantitative EDS analysis are presented in Table I. The amount of zinc incorporated into each coating was correlated to the amount of DC power applied to the zinc target, indicating that the zinc concentration within the coating can be selectively controlled.

Table I: Composition of films produced with differing parameters.

	Power Applied to Zn Target	Wt. % Zn
HA	-	0
Low Zn	1.5 W	3.4
High Zn	3.0 W	6.4

Stylus profilometry indicated the films were approximately 600 nm in thickness. The corresponding rate of deposition (~150 nm/hr) is consistent with previously reported results [7]. SEM indicated that the as-deposited films followed the surface morphology of the Ti-6Al-4V substrate. Following heat treatment, the films retained a similar morphology, but there was an infrequent appearance of small cracks in the film surface. The incorporation of zinc appeared to have little effect on film morphology.

The results of XRD are presented in Figures 1 and 2. As-deposited films were amorphous in nature, but annealed films showed a crystalline apatite structure. In high Zn coatings, a preferred orientation with high intensity (h 0 0) peaks was observed, suggesting that zinc incorporation may affect the orientation of crystal growth during the annealing process.

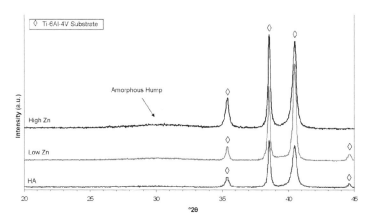

Figure 1: XRD patterns of as-deposited films.

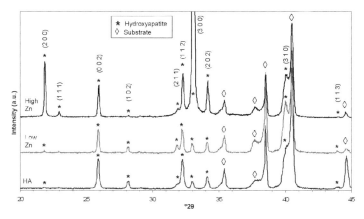

Figure 2: XRD patterns of films following heat treatment at 600°C for 3 hours.

In Vitro Analysis

The results of the AlamarBlue™ and CyQUANT™ assays for cell growth are presented in Figures 3 and 4, respectively. While initial cell attachment and growth was similar for all coatings, after 7 days cells on low Zn coatings showed significantly more proliferation than cells on both HA and high Zn surfaces. After 14 days, both AlamarBlue™ and CyQUANT™ confirmed a significantly greater number of cells on low Zn films than on HA films.

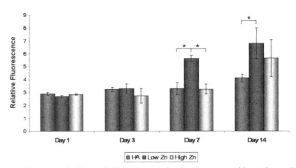

Figure 3: Cell metabolic activity over 14 days as measured by AlamarBlue™ assay (*p≤0.05).

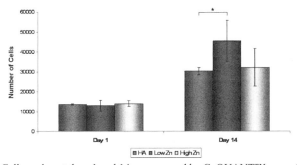

Figure 4: Cell number at days 1 and 14 as measured by CyQUANT™ assay (*p≤0.05).

Fluorescent microscopy showed that cells attached well and spread on all surfaces. Typical cell morphologies for HOBs were observed (Figure 5). Cells seemed to orient themselves along the grooves created during the grinding of substrates.

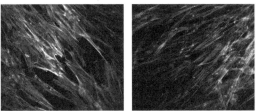

Figure 5: Fluorescent microscopy images of HOBs on low Zn coatings at day 3 (left) and day 14 (right).

CONCLUSIONS

Zinc-doped hydroxyapatite thin films on Ti-6Al-4V substrates have been produced using a magnetron co-sputtering process. Coatings produced with a DC power applied to the zinc target at 1.5 W and 3.0 W resulted in the incorporation of 3.4 wt. % zinc and 6.4 wt. % zinc into the film, respectively. XRD confirmed the crystallization of the apatite coating during heat treatment at 600° C in a moist argon atmosphere. *In vitro* analysis using a HOB model showed that the inclusion of 3.4 wt. % zinc into sputtered hydroxyapatite films enhances osteoblast proliferation and activity. Magnetron-sputtered zinc-doped hydroxyapatite thin films thus show great promise as coatings on load-bearing implants for orthopedic applications.

ACKNOWLEDGMENTS

Funding for this work was provided by the Furlong Research Charitable Foundation, St. John's College, and the Cambridge Overseas Trust. Funding to present this work was generously provided by St. John's College and the Armourers and Brasiers' Company.

REFERENCES

1. L.L. Hench, J. Am. Ceram. Soc. **74**, 1487 (1991).
2. K. de Groot, R.G.T. Geesink, C.P.A.T. Klein, and P. Serekian, J. Biomed. Mater.Res. **21**, 1375 (1987).
3. K. Søballe, E.S. Hansen, H. Brockstedt-Rasmussen, and C. Bunger, J. Bone Joint Surg.**75B**, 270 (1993).
4. I. Kanasniemi, C. Verheyen, E. van der Velde, and K. de Groot, J. Biomed. Mater. Res. **28**, 563 (1994).
5. J. Dalton and S. Cook, J. Biomed. Mater. Res. **29**, 239 (1995).
6. C. Massaro, M. Baker, F. Cosentino, P. A. Ramires, S. Klose, E. Milella, J. Biomed. Mater. Res. **58**, 651 (2001).
7. J. E. G. Hulshoff, K. van Dijk, J. P. C. M. van der Waerden, J. G. C. Wolke, W. Kalk, and J. A. Jansen, J. Biomed. Mater. Res. **31**, 329 (1996).
8. E.S. Thian, J. Huang, S.M. Best, Z.H. Barber and W. Bonfield, Materials Science and Engineering: C, **27**(2), 251 (2007).
9. B. Moonga and D. Dempster, J. Bone and Min. Res. **10**, 453 (1995).
10. A. Itoa, H. Kawamurab, M. Otsukac, M. Ikeuchia, H. Ohgushia, K. Ishikawad, K. Onumaa, N. Kanzakia, Y. Sogoa, and N. Ichinose, Materials Science and Engineering: C **22**, 21 (2002).

Mater. Res. Soc. Symp. Proc. Vol. 1140 © 2009 Materials Research Society   1140-HH05-09

Active Cooperative Assemblies Towards Nanocomposites

M. S. Toprak,* Carmen Vogt and Abhilash Sugunan
Department of Microelectronics and Applied Physics, Division of Functional Materials, Royal
Institute of Technology, SE16440, Stockholm Sweden
*Correspondance Author : Tel:+46 8 7908344 Fax:+46 8 7909072 e-mail: toprak@kth.se

ABSTRACT

This work reports on the fabrication of novel type of assemblies bearing magnetic nanoparticles and inorganic shells prepared via a biomimetic route of complex coacervation. Magnetic nanoparticles fabricated under controlled conditions were surface modified with poly-acrylic acid (PAA). Subsequently, PAA spontaneously formed spherical assemblies in contact with certain ions, such as Ca^{2+}. The stability of these microspheres against environmental alterations such as pH, ionic strength, and dilution was increased through cross-linking. Ethylene diammine (EDA) was used as a cross-linker, which resulted in mechanically stabilized system that does not show sensitivity towards the external pH values. Important parameters for the formation of these coacervates as well as mechanism of formation and cross-linking have been evaluated by FTIR analysis. The cooperative assemblies are still active for further reaction and were used for the growth of an inorganic aluminum oxide shell. SEM analysis of these spheres showed that the structures are hollow with a large interior volume. A biocompatible outer surface combined with the magnetic functionality is very important for the targeted drug delivery devices for biomedical applications.

INTRODUCTION

Nano- and microscale hollow spherical materials have attracted considerable attention in the last decade and have been extensively studied for possible practical applications in materials science and medicine.[1,2,3,4] Both inorganic and polymeric hollow microspheres have been reported in the literature. Among them, hollow particles made from metal (e.g., gold),[5] metal oxides (e.g., Al_2O_3, TiO_2, ZrO_2),[6] silica,[7] composites (e.g., ZnS, CdS),[8,9] and polymers (e.g., poly(methylmethacrylate),[10] poly(N-isopropylacrylamide),[11] polyorganosiloxane,[12] poly (acrylamide)/poly(acrylic acid) (PAAM/PAAC),[13] poly(styrene),[14] poly(3,4-ethylenedioxythio-phene) (PEDOT),[15] polyaniline (PANI)), polypyrrole (PPY),[16] have been fabricated with various diameters and wall thicknesses.

Methods for the preparation of core-shell microspheres reported in the literature generally involve either physicochemical or chemical processes. In the physicochemical process, an organic or inorganic substance is precipitated at the core surface through adsorption by means of electrostatic or chemical interactions or during solvent evaporation. On the other hand, fabrication of core-shell particles by chemical processes involves preparation of various seeds (templates),[17,18] followed by various multistep polymerization reactions,[11,12] such as emulsion,[19] microemulsion, or suspension.[14] Subsequently, calcination[4] or solvent etching[20] is used to remove the template materials. In most cases, however, the formation of a uniform shell surrounding the core, as well as control of the shell thickness is difficult to achieve because the polymerization is not restricted to the surface of the templates. A report by Wong and co-workers demonstrated the formation of supramolecular aggregates between cationic polyamines and multivalent counteranions via ionic crosslinking in a two-step process; negatively charged nanoparticles deposit around these aggregates to form a multilayer-thick shell.[21] Complex coacervation is a natural phenomenon that presents new opportunities for single-step syntheses of hybrid materials that are composed from predefined nanoscale objects, and can be defined as a spontaneous aqueous phase separation, in which liquid-like microspheres are produced from oppositely charged chemical entities.[22] In our earlier work we reported on the cooperative assembly of citrate coated magnetic nanoparticles with polyamines.[23],[24] In this work, a very mild, template-free method, based on spontaneous cooperative assembly of PAA coated magnetic nanoparticles with +2 charged ions for the fabrication of nanocomposite hollow microspheres is presented.

EXPERIMENT

Chemicals

$FeCl_2.4H_2O$, $FeCl_3.6H_2O$, NH_4OH, HCl, Poly acrylic acid, (PAA, M_w: 200), PAA Na salt (M_w: 15,000) ethylenediamine (EDA, Merck). All chemicals were used as received and aqueous solutions were prepared by dissolving the corresponding chemicals in DI water, 18 $M\Omega$.

PAA modified magnetic nanoparticles.

Superparamagnetic iron oxide nanoparticles were prepared by using a previously described procedure.[25] In a typical process, a mixture of Fe^{2+} and Fe^{3+} were hydrolyzed by the addition of NH_3 solution at pH >10 in an oxygen-free atmosphere. Afterwards, the reaction mixture was heated up to 80 °C under Ar flow, followed by the addition of PAA. Subsequently, the reaction mixture was cooled down and surface modified magnetic nanoparticles (MNPs) were collected by magnetic decantation.

Formation of Hybrid Coacervates

A solution of magnetic nanoparticles (MNP) having a net negative surface charge due to PAA was mixed with Ca^{2+} solution, upon which the solution turned cloudy. A typical sample was fabricated by mixing 20 μL of 2 mg/mL Ca solution with 120 μL MNP (1.6 mg/ml PAA). The reaction was mixed vigorously for 15 seconds using a vortex-mixer.

Alumina coating

A colloidal alumina solution was added dropwise into coacervate solution and alumina nanoparticles were adsorbed by the oppositely charged microsphere surface forming thick shells. Samples were then embedded analyzed under SEM.

Iron oxide nanoparticles, MNPs, made by aqueous precipitation were analyzed by TEM and a typical batch is represented by TEM image in Fig 1(a). Magnetic evaluation of these nanoparticles revealed a saturation magnetization of 53 emu/g with no remanent magnetization, revealing superparamagnetic character. MNPs were then surface functionalized with PAA at

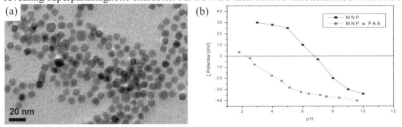

Figure 1. (a) TEM image of as-prepared and surface functionalized magnetic iron oxide nanoparticles; (b) Zeta potential analysis of as-prepared and surface functionalized MNPs.

elevated temperature. The success of modification was monitored by zeta potential analysis and FTIR analysis. Zeta potential analysis of bare MNPs before and after surface modification is presented in Fig. 1(b). The image reveals that as-made MNPs exhibit a point-of-zero charge at pH ~7 which shifts to pH ~2 after PAA adsorption due to the carboxylic acid moities expressed on the PAA molecule. A further confirmation of the PAA adsorption is via FTIR analysis as shown in Fig. 2(a). Typical FTIR bands of PAA is exhibited in the FTIR spectrum of MNPs activated with PAA. The most striking change is the dramatic intensity reduction of the band at ~1700 cm^{-1} representing C=O in PAA for the PAA coated MNPs. This suggests that PAA could be in bidentate chelating or bridging mode to magnetite nanoparticle surface thereby generating O-C-O stretching vibrations at ~1520 cm^{-1}, as schematically described in Fig. 2(b). Possible adsorption schemes of PAA onto magnetite surface as monodentate, bidentate chelating and bidentate bridging are summarized in Fig. 2(c) which give rise to variations of vibration bands

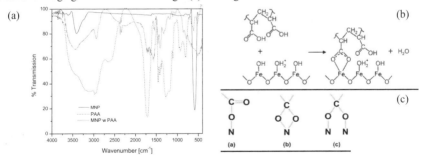

Figure 2. (a) FTIR analysis of as-made MNPs, PAA and MNPs with PAA. (b) Adsorption scheme I (chemical formula) and (c) adsorption scheme II with the structures (a), (b) and (c), where **N** represents Fe metal center.

and their separation in the FTIR spectrum.

Following surface modification, MNPs with PAA were reacted with Ca^{2+} ions. As a result of this electrostatic interaction of PAA with Ca^{2+} a spontaneous cooperative assembly took place forming spherical features. These features were observed under optical microscope (Fig 3(a) exhibited a concentrated brown color at the periphery of the spheres, indicating the accumulation of MNPs at the walls. Upon drying the observed features disappeared due to the destabilization of the electrostatic interaction. Fluorescence image of the assembly using FITC dye indicated that these spheres were hollow as shown in Fig. 3(b). Size of these nanocomposite spheres were found to be dependant on the ratio of n_{PAA}/n_{Ca}^{2+} as summarized in Figure 3(c).

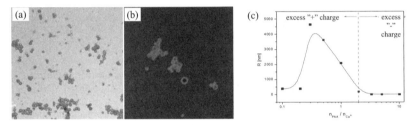

Figure 3. (a) Optical and (b) fluorescence micrograph of cooperatively assembled of Ca^{2+} - MNP w/ PAA; (c) Size of microspheres as a function of PAA/Ca^{2+} ratio

In order to stabilize these cooperative assemblies EDA was introduced into the solution containing nanocomposite microspheres. Optical microscopy revealed that the spheres were intact even after the solvent dried out. Amount and addition rate of cross-linker EDA was found to be very critical on the size and morphology of cross-linked spheres. Figure 4 shows SEM micrographs for some of the cross-linked samples with various quantities of EDA introduced to the reaction solution at different rates – for the case nPAA/nCa=1. When amount of EDA was

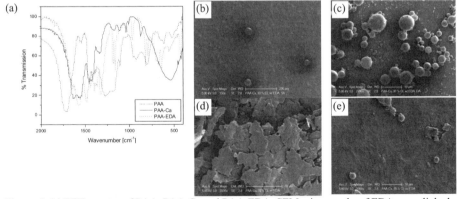

Figure 4. (a) FTIR spectra of PAA, PAA-Ca and PAA-EDA; SEM micrographs of EDA cross-linked microsphere samples. EDA addition was performed at following concentrations and rates: (b) 30 wt-% EDA, slow addition; (c) 30 wt-% EDA, fast addition; (d) 70 wt-% EDA, slow addition; (a) 70 wt-% FDA, fast addition. n_{PAA}/n_{Ca} 1

adjusted to 30 wt-% of the PAA concentration, fast addition of EDA produced very large spheres of about ~40 µm in size; Fig. 4(b). When the addition was made slowly much smaller spheres in the range 500 nm – 8 µm were obtained; Fig. 4 (c). For the case where amount of EDA was adjusted to 70 wt-% of the PAA concentration, slow addition of EDA produced sheet like polymeric aggregates (Fig. 4(d)) while fast addition produced excessive aggregates and spheres with average size of 3 µm (Fig. 4(e)).

The mechanism of cross-linking was evaluated by FTIR analysis and results are presented in Figure 4(a). According to FTIR analysis results, conjugation of PAA to EDA is similar that of MNPs. The typical sign is the disappearance of the C=O peak at 1700 cm^{-1} upon interaction with EDA and/or Ca^{2+}. That indicates the interaction of PAA should be in bidentate chelating or bridging mode in the coacervate and cross-linked form. Appearance of new bands at 1632 and 1556 cm^{-1} upon Ca^{2+} addition is due to the bidentate interaction of PAA w/ Ca^{2+}-COO$^-$ …Ca^{2+}… COO$^-$ bidentate chelating. Important absorption bands for EDA at ~1640-1670 cm^{-1} for N-H bending and 860-890 cm^{-1} for primary amine N-H out-of-plane bending vibrations dramatically decreased in intensity or disappeared completely due to the interaction of amine ends with the carboxyl groups of PAA.

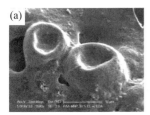 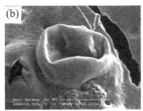 

Figure 5. SEM micrographs of (a-b) cross-linked MNP w/PAA/Ca^{2+} spheres collapsed due to the high vacuum in sample preparation or SEM analysis; (c) Aluminum oxide nanoparticles are deposited around these assemblies showing the preserved hollow structure.

After cross-linking SEM analysis was performed on the MNP w/PAA/Ca^{2+} spheres. The structures observed were generally burst or collapsed spheres under high vacuum conditions during SEM sample preparation or SEM analysis; Fig. 5(a-b). The outer surface of these assemblies was found to be charged and this was used to assemble an inorganic shell layer around the spheres. A resulting structure is presented in Fig. 5(c) which shows that both the shape and the hollowness are preserved.

CONCLUSIONS

In this work we presented a biomimetic route for a cooperative assembly of PAA grafted nanoparticles into microspheres through complexation with Ca^{2+} ions. The method presented is novel as it creates functional nanoparticles embedded nanocomposites in one step without using any colloidal template. Thus nanoparticles can carry both magnetic functionality and an outer surface that is biocompatible. As such spontaneously organized nanoparticles assemblies were then stabilized by cross-linking the carboxyl groups using EDA as cross-linker. Detailed FTIR analysis revealed the interaction between the different components in the system. Ratio of components is found to be a very important parameter that is effective in controlling the size of these formations. The cooperative assemblies are still active for further reaction and were used for the growth of an inorganic aluminum oxide shell around the microspheres. These capsules with a large interior volume and biocompatible outer surface combined with the magnetic functionality are very important potential drug delivery devices for biomedical applications. Our findings enable us to design bioinspired biocompatible hybrid composites that may contribute significantly to the future design and fabrication of bioinspired advanced functional materials.

ACKNOWLEDGMENTS

This work is partially supported by FP6 Program EU Projects Nanoear (NMP4-CT-2006-026556) and Biodiagnostics (NMP4-CT-2005-017002). The fellowship from Knut and Alice Wallenbergs Foundation (MST) is thankfully acknowledged (No:UAW2004.0224).

REFERENCES

1 P. K. Naraharisetti, M. D. N. Lew, Y. C. Fu, D. J. Lee, C. H. Wang, *J. Controlled Release* **2005**, *102*, 345.

2 K. J. Pekarek, J. S. Jacob, E. Mathiowitz, *Nature* **1994**, *367*, 258.

3 A. D. Skelhorn, *U. S. Patent 2005126441*, **2005**.

4 W. Li, X. Sha,W. Dong, Z.Wang, *Chem. Commun.* **2002**, 2434.

5 S. Chah, J. H. Fendler, J. Yi, *J. Colloid Interface Sci.* **2002**, *250*, 142.

6 T. C. Chou, T. R. Ling, M. C. Yang, C. C. Liu, *Mater. Sci. Eng. A* **2003**, *359*, 24.

7 E. Bae, S. Chah, J. Yi, *J. Colloid Interface Sci.* **2000**, *230*, 367.

8 A. Wolosiuk, O. Armagan, P. V. Braun, *J. Am. Chem. Soc.* **2005**, *127*, 16 356.

9 J. S. Hu, Y. G. Guo, H. P. Liang, L. J. Wan, C. L. Bai, Y. G. Wang, *J. Phys. Chem. B* **2004**, *108*, 9734.

10 M. A. Winnik, C. L. Zhao, O. Shaffer, R. R. Shivers, *Langmuir* **1993**, *9*, 2053.

11 Q. Sun, Y. Deng, *J. Am. Chem. Soc.* **2005**, *127*, 8274.

12 S. Mishima, M. Kawamura, S. Matsukawa, T. Nakajima, *Chem. Lett.* **2002**, 1092.

13 X. C. Xiao, L. Y. Chu, W. M. Chen, J. H. Zhu, *Polymer* **2005**, *46*, 3199.

14 G. H. Ma, A. Y. Chen, Z. G. Su, S. Omi, *J. Appl. Polym. Sci.* **2003**, *87*, 244.

15 M. G. Han, S. H. Foulger, *Chem. Commun.* **2004**, 2154.

16 Y. Yang, Y. Chu, F. Yang, Y. Zhang, *Mater. Chem. Phys.* **2005**, *92*, 164.

17 S. Fujikawa, T. Kunitake, *Langmuir* **2003**, *19*, 6545.

18 F. Caruso, R. A. Caruso, H. Möhwald, *Science* **1998**, *282*, 1111.

19 K. Zhang, H. Chen, X. Chen, Z. Chen, Z. Cui, B. Yang, *Macromol. Mater. Eng.* **2003**, *288*, 380.

20 T. K. Mandal, M. S. Fleming, D. R. Walt, *Chem. Mater.* **2000**, *12*, 3481.

21 J. N. Cha, M. H. Bartl, A. Popitsch, M. S. Wong, T. J. Deming, G. D. Stucky, *Nano Lett.* **2003**, *3*, 907.

22 H. G. Bungenberg de Jong, in Colloid Science; Vol. II, (Ed. H. R. Kruyt), Elsevier, Amsterdam, 1949, Ch. 8, pp. 232–258.

23 M. S. Toprak, B. J. McKenna, J.H. Waite, G.D. Stucky, *Adv. Mater.* **2007**, *19*, 1299.

24 M. S. Toprak, B. J. McKenna, J.H. Waite, G.D. Stucky, *Chem. Mater.* **2007***, 19,* 4263-4269.

25 F. A. Turinho, R. Franck, R. Massart, *Prog. Colloid Polym. Sci.* **1989**, *79,* 128.

Mater. Res. Soc. Symp. Proc. Vol. 1140 © 2009 Materials Research Society   1140-HH06-35-DD03-35

Poly(vinylidene fluoride) Electrospun Fibers for Electroactive Scaffold Aplications: Influence of the Applied Voltage on Morphology and Polymorphism

V. Sencadas[1,2], J.C. Rodríguez Hernández[2], C. Ribeiro[1], J.L. Gómez Ribelles[2, 3,4] and S. Lanceros-Mendez[1]

[1] Dept. de Física, Universidade do Minho, 4710-057 Braga, Portugal
[2] Centro de Biomateriales e Ingeniería Tisular, Universidad Politécnica de Valencia, 46022 Valencia, Spain
[3] Regenerative Medicine Unit. Centro de Investigación Príncipe Felipe, Autopista del Saler 16, 46013 Valencia, Spain
[4] CIBER en Bioingeniería, Biomateriales y Nanomedicina, Valencia, Spain

ABSTRACT

Electrospinning is a well known method for processing flexible and highly porous nanofiber scaffolds by applying a high electric field to a droplet of polymer solution or melt.

The morphology of the electrospun Poly(vinylidene floride) -based scaffolds (EPS) was investigated by scanning electron microscopy (SEM). In particular, the average fiber diameter decreases with increasing applied voltage. The samples are mainly in the electroactive β-phase, the amount of α-phase decreasing with increasing processing voltage. Thermal properties and polymorphism of the polymeric scaffolds were investigated by differential scanning calorimetry and infrared spectroscopy, respectively.

INTRODUCTION

Poly(vinylidene fluoride), PVDF, is known for its chemical resistance and for its electroactive properties.In particular ferro-, piezo- and pyroelectricity show outstanding values for polymer systems[1].

PVDF is a semi-crystalline polymer which shows an unusual polymorphism among polymers, showing at least four crystalline phases known as α, β, γ, δ. The one with the best piezoelectric and pyroelectric properties is the β-phase. The electroactive properties heavily depend on the phase content, microstructure and degree of crystallinity of the samples, which in turn depend on the processing conditions [1].

Electrospinning is an easy and efficient method to prepare submicron to nano-scale fibers [2]. PVDF membranes were prepared using electrospinning for various applications [3-6]. However, there is little information about the controlling of the crystalline phase of the polymer in electrospun PVDF.

In the present work the influence of the applied voltage during electrospinning process on the diameter of the PVDF nanofibers, crystalline phase and thermal stability of the polymeric scaffolds were investigated.

EXPERIMENTAL

Materials: Poly(vinylidene fluoride) (Solef 1100) was obtained from *Solvay*. N,N-Dimethyl Formamide (DMF) was purchased from *Merck*. The polymer was dissolved in a solution of DMF with a concentration of 20% (w/w) of PVDF. PVDF was dissolved at room temperature in a magnetic stirrer. The viscosity of the solution was measured in a rheometric apparatus *ViscoStar L plus*.

Electrospinning: The polymer solution was placed in a plastic syringe (5ml) fitted with a a steel needle of tip-diameter 250 μm. Electrospinning was conducted in a range of 15 o 30 kV with a high voltage power supply from *Glassman* (model PS/FC30P04). A syringe pump (from *Syringepump*) was used to feed the polymer solutions into the needle tip at rate of 4ml/h. The electrospun fibers were collected in an aluminium plate that was placed at 15cm from the needle.

Characterization: Electrospun fibers were coated with gold using a sputter coating and their morphology was observed by SEM (model JSM-6300, JEOL) with an accelerating voltage of 20kV and a magnification of 1,000-10,000x. The size of nanofibers was measured on 4,000x magnified SEM images using the Scion Imaging software (*Scioncorp*).

Infrared measurements (FTIR) were performed at room temperature in a *Perkin-Elmer Spectrum 100* apparatus in ATR mode from 4000 to 650cm^{-1}. Differential scanning calorimetry measurents (DSC) were performed in a *Mettler-Toledo* DSC823e apparatus at a heating rate of 10°C.min^{-1}. The samples for the DSC studies were cut into small pieces from the middle region of the electrospun membranes and placed into 40μl aluminium pans. All experiments were performed under a nitrogen purge.

RESULTS AND DISCUSSION

The morphology of the electrospun polymer fibers is dependent on many parameters such as solution properties, processing and ambient conditions. Therefore, solution properties are one of the most important properties parameters in electrospinning processing technique. Molecular weight of the polymer is one of the first issues to be considered. The molecules of the polymer should be long enough to entangle. Other important issue is the viscosity of the polymer solution. If the viscosity is low, although the polymer molecules are entangled, the molecular chains are too separated from each other and as a result, fibers are not formed and droplets, known as beads, are produced instead due to the breaking of the jet. On the other hand, electrospinning is suppressed at too high polymer concentrations because it prohibits the continuous flow of the solution to the tip [7].

In this work, solutions of PVDF with concentrations lower than 20% were impossible to electrospin due to the low viscosity of the solutions. The results obtained for lower concentrations suggest that the entanglement of the molecules was not strong enough to overcome the repulsion of positive charge arising from the applied external voltage. Solutions with 20% of PVDF were successfully electrospun on aluminized flat plates with voltages ranging from 15 to 30kV and a traveling distance of 15cm. The viscosity for this solution was 831.91 cP and perfect fibers were formed.

Figure 1 show the SEM images of typical electrospun PVDF fibers obtained at a travel distance of 15cm and at 25kV.

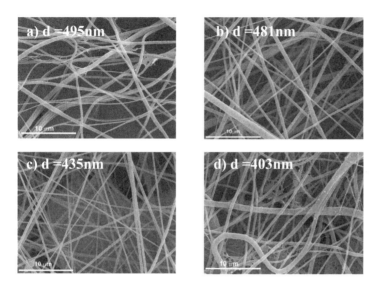

Figure 1. SEM image of PVDF electrospun nano fibers obtained from a solution of 20/80 (20% PVDF + 80%DMF w/w) at a traveling distance of 15cm.

Average fiber diamenter (AFD) was calculated from the images and is shown in Figure 1. It can be observed that the AFD of the polymeric electrospun scaffolds decrease from 495nm to 403nm with increasing voltage from 15kV to 30kV. In electrospinning, the high voltage is the drive of the electrospinning processing. The formation of ultra fine fibers is mainly achieved by the stretching and acceleration of the jets in a high electric field [8, 9]. High applied voltages can result in a higher charge density on the surfaced of the ejected jets, thus the jet velocity increases and higher elongation forces are imposed to the jet. Consequently, the diameter of the final fibers becomes gradually smaller with increasing applied voltages [10]. At the same time, an increase in the applied voltage also enhances the degree of the instability of the jets during the travel from the needle tip to the metal collector, which result in a broader distribution of the fiber diameters [10].

There are several factors influencing the crystalline phase present in electrospun PVDF fibers during the electrospinning process, such as solvent concentration, electrospinning temperature, feeding rate and distance from the tip to the collector. In the present work several voltages were applied in order to obtain electrospun PVDF fibers. The influence of the voltage on the crystalline phase of the polymer was characterized by infrared FTIR spectroscopy. Figure 2 shows the FTIR spectra obtained for the different electrospun scaffolds.

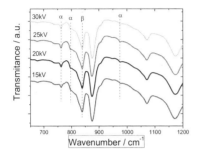 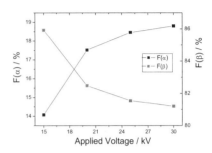

Figure 2. a) Detail of the FTIR spectra for the PVDF electrospun scaffolds obtained at different applied voltages (lines indicate the absorption bands characteristics of the α and β-phase), **b)** Amount of α and β-phase within the crystalline phase of the polymeric scaffolds.

Figure 2a shows the characteristic absorption modes for the α-phase (855, 766, 615 and 531 cm^{-1}) and β-phase (510 and 840 cm^{-1}) without traces of the γ-phase (431, 776, 812 and 833 cm^{-1}) [1, 11].

The general appearance of the spectra is similar for all samples: neither vibrational modes are totally suppressed nor new modes seem to appear due the increasing of the applied voltage. The evolution of the phase with applied voltage was also studied by FTIR on these samples. The method explained elsewhere [11] was applied to calculate the relative amount of α and β-phase present in the different samples (equation 1).

$$F(\beta) = \frac{A_\beta}{(K_\beta / K_\alpha)A_\alpha + A_\beta} \quad (1)$$

Here, F(β) represents the β-phase content; A_α and A_β the absorvancies at 766 and 840 cm^{-1}, corresponding to the α and β-phase of the material; K_α and K_β are the absorption coefficient at the respective wavenumber. The value of K_α and K_β are 6.1×10^4 and 7.7×10^4 cm^2.mol^{-1}, respectively [11]. The results obtained for the non-polar α-phase and for the electroactive β-phase amount present in each sample is illustrated in Figure 2b.

Figure 2b demonstrates that the crystalline main phase present in the polymeric scaffolds is the β-phase, and the amount of the electroactive phase increases with increasing the applied electrical voltage. The maximum of β-phase present in the scaffolds is ≅ 86%, obtained for V = 30kV. On the other hand, it was observed that the amount of the non-polar α-phase decreases with increasing voltage and a minimum of ≅ 14%, was obtained for the sample processed at 30kV.

In order to determine possible modifications in crystal structure and melting behavior, DSC measurements were performed on raw PVDF and in the electrospun scaffolds with different average fiber diameter. Figure 3a shows the DSC thermograms of the samples. All samples showed similar endothermic peaks. It is to notice that the DSC curves show two melting peaks.

The one that occurs at lower temperatures is related to the melting of the crystalline fraction of the α-phase present in the sample and the one at higher temperatures is associated to the melting of the β-phase of the polymer. These results are in accordance with the FTIR spectra of the samples what also demonstrate the co-existence of the α and β crystalline phase in the PVDF electrospun samples.

The degree of crystallinity (ΔXc) of each sample was determined from its DSC curves using equation 2:

$$\Delta X_c = \frac{\Delta H}{x\Delta H_\alpha + y\Delta H_\beta} \tag{2}$$

where, ΔH is the melting enthalpy of the sample under consideration; ΔH_α and ΔH_β are the melting enthalpies of a100% crystalline sample in the α and β-phase respectively and x and y are the amount of the α and β-phase present in the, respectively. In this study a value of 93.07 Jg^{-1} and 103.4 Jg^{-1} was used for the ΔH_α and ΔH_β, respectively [2, 11].

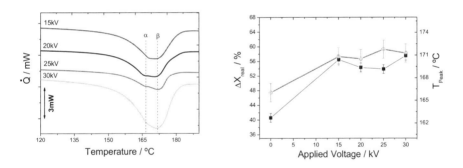

Figure 3. a) DSC thermograms for the electrospun PVDF scaffolds processed at different applied voltages, **b)** Influence of the applied voltage in the degree of crystallinaty and melting peak temperature of the electrospun scaffolds. The values for 0 voltage correspond to β-PVDF films prepared by stretching [11].

Compared to β-PVDF films prepared by the conventional stretching method [1, 11], all electrospun samples show higher crystallinity content, corresponding to higher melting enthalpy. The solidification of the polymer molecular chains under high elongational rate during the spinning process may enhance the development of the polymer crystallinity. In other words, the molecular chains of PVDF have time to crystallize before they are immoblized in the collector.

The peak temperature of the DSC thermograms (Figure 3b), also show that samples of PVDF obtained by electrospinning have higher melting temperature than the PVDF films.

Melting temperature and crystallinity of the electrospun samples on the other hand does not suffer significant changes with the applied voltage.

CONCLUSIONS

Poly(vinylidene fluoride) electrospun nano fibers has been successfully obtained. The samples processed by electrospinning are mainly in electroactive β-phase but a small amount of the non polar α-phase still remains. . The α-phase decreases with increasing applied voltage between the tip of the needle and the metal collector.

The degree of crystallinity and the melting temperature of the electrospun fibers are higher than for β-PVDF films.

ACKNOWLEDGMENTS

The authors thank the Portuguese Foundation for Science and Technology (FCT) (Grants PTDC/CTM/69362/2006 and PTDC/CTM/69316/2006) and INL project 156: SIMBIO). V. Sencadas thanks the FCT for the PhD Grant (SFRH/BD/16543/2004) C. Ribeiro thanks the INL for the PhD Grant. and JLGR acknowledge the support of the Spanish Ministry of Science through project No. MAT2007-66759-C03-01 (including the FEDER financial support). The authors also thank to *Solvay* for providing the excellent quality material and also to the UPV Microscopy Service for the useful help.

REFERENCES

1. A. J. Lovinger, In: Developments in Crystalline Polymers; Basset, D.C., Ed. Applied Science Publications: London, 1981.
2. J. Doshi, D. H. Reneker, J. Electrost. 35, 151 (1995).

3. S.W. Choi, J. R. Kim, Y. R. Ahn, S.M. Jo, E. J. Cairns, Chem.Mater. 19, 104 (2007).
4. S. S. Choi, Y. S. Lee, C. W. Joo, S. G. Lee, J. K. Park, K. S. Han, Electrochim. Acta 50, 339 (2004).
5. K. Gao, X. G. Hu, C. S. Dai, T. F. Yi, Mater. Sci. Eng. B 131, 100 (2006).
6. R. Gopal, S. Kaur, Z. W. Ma, C. Chan, S. Ramakrishna, T. Matsuura, J. Membr. Sci., 281, 581 (2006).
7. K. H. Lee, H. Y. Kim, H. J. Bang, Y. H. Jung, S. G. Lee, Polymer 44, 4029 (2003).
8. D.H. Reneker, A.L. Yarin, H. Fong, S. Koombhonge, J. Appl. Phys. 87, 4531 (2000).
9. X.H. Qin, Y.Q. Wan, J.H. He, J. Zhang, J.Y. Yu, S.Y. Wang, Polymer 45, 6409 (2004).
10. K. Gao, X. Hu, C. Dai, T. Yi, Materials Science & Engineering B 131, 100 (2006).
11. V. Sencadas, V. M. Moreira, S. Lanceros-Mendez, A. S. Pouzada, R. Gregorio Jr., Mat. Sci. Forum, 514-16, 872 (2006).

Optimised Synthetic Route for Tuneable Shell SiO$_2$@Fe$_3$O$_4$ Core-Shell Nanoparticles

Carmen M. Vogt[1], Muhammet Toprak[1], Jingwen Shi[2], Neus Feliu Torres[2], Bengt Fadeel[2],
Sophie Laurent[3], Jean-Luc Bridot[3], Robert N. Müller[3], Mamoun Muhammed[1]
[1]Functional Materials, Royal Institute of Technology, Isafjordsgatan 22, 16440 Kista, Sweden
[2]Institute of Environmental Medicine, Karolinska Institutet, Nobels väg 13, S-171 77
Stockholm, Sweden
[3]Department of General, Organic, and Biomedical Chemistry, NMR and Molecular Imaging
Laboratory, University of Mons-Hainaut, B-7000 Mons, Belgium

ABSTRACT

Multifunctional nanoparticles (that have in their structure different components that
can perform various functions) are subject of intensive research activities as they find a large
variety of applications in numerous biomedical fields from enhancement of image contrast in
MRI to different magnetically controllable drug delivery systems.

In this study we report on the synthesis of well-separated, monodisperse single core-
shell SiO2@Fe3O4 nanoparticles with an overall diameter of ~30 nm. The influence of
stirring rate and reaction time on synthesis of tuneable shell thickness core-shell nanoparticles
is reported. Particles' cell toxicity and performance as MRI contrast agents were also studied
due to their promising biological applications (as contrast agents, cell labelling and
separation, drug delivery systems, etc.) and results are promising in terms of MRI
performance as well as having no significant cytotoxicity.

INTRODUCTION

The development of new advanced applications in biomedical related fields like drug
delivery, diagnostics and biomolecules separations and detection systems introduced the
demand for a new generation of nanoparticles called multifunctional nanoparticles. These are
complex, well controlled nanoparticle architectures that carry several components with
different functionality. The core – shell are one type of multifunctional nanoparticles carrying
the magnetic functionality in the magnetic core, the surface modification or drug loading
capacity in the shell.

For applications in biology and medical diagnostics and therapy the
superparamagnetic (no magnetic remanence) iron oxide nanoparticles (SPION) are
desired[1,2]. These iron oxides based nanoparticles with a narrow size distribution and
superparamagnetic character are FDA approved materials for biological applications. Several
synthetic routes have been investigated for production of magnetic nanoparticles[1,3,4,5]. The
thermal decomposition route based on iron oleate as precursor is a technically easy, nontoxic
and environmentally friendly route to produce high yield of iron oxide nanoparticles with a
finely tailored size from 5 nm to 22 nm[6].

Surface modification of SPION iron oxide nanoparticles is a necessity in order to
avoid agglomeration but also to increase the biocompatibility and the retention time in the
blood stream. Silica is one of the preferable materials for surface coating when high
biocompatibility, stability and increase in residence time are desired.

As shown by the early work of Stober[7], under basic conditions TEOS undergoes
hydrolysis and polycondensation reactions which result in the formation of monodisperse
spherical particles of amorphous silica. The water – in – oil microemulsion called also inverse

microemulsion takes these reactants in a confined space (the water droplets) resulting highly monodisperse, very well controlled size silica nanoparticles. Making silica nanoparticles by reverse micelle media possesses the potential to effectively control the size and morphology of the nanomaterials. The influence of various factors on particles size, polydispersity and morphology of the silica nanoparticles was studied[8,9].

The water-in-oil, or inverse, microemulsion is one versatile way of obtaining the multifunctional, well defined structures, such as core – shell nanoparticles, where the core has magnetic properties[10,11,12].

In this work, we report on production of monodispersed size exclusively single core Fe_3O_4 – silica shell nanoparticles with adjustable (up to ~13 nm) silica shell thickness (with the potential to further increase the shell thickness).

THEORY

Chemicals

Iron oxide hydrated, FeO(OH), Triton-X100, cyclohexane and hexanol were purchased from Sigma Aldrich and hexane, oleic acid, tetraethyl orthosilicate (TEOS) and NH_4OH (28%) from Fluka. All chemicals were used as received without further purification. The water used was MilliQ pure water with a resistivity of 18 MΩ and ethanol was of 99,9 % purity. The microemulsion formation was done by mechanical stirring with a centrifugal stirrer of 40 mm diameter of the paddles purchased from VWR.

Transmission electron microscopy (TEM) analysis of the nanoparticles' size and morphology were performed with JEOL 2100 high resolution FEG-TEM unit at 200 kV acceleration voltage.

Atomic Absorption Spectroscopy (AAS) for concentration determination was performed with Varian SpectrAA-220.

The magnetic measurements were performed using a vibrating sample magnetometer, VSM (NUOVO MOLSPIN, UK) in the magnetic field range of ±1 Tesla.

The cell toxicity tests 3-(4,5- dimethyldiazol-2-yl)-2,5 diphenyl Tetrazolium Bromid (MTT) assay were performed on A549 cell line culture.

Synthesis of magnetic core

The magnetic nanoparticles were prepared by thermal decomposition of iron oleate in a high boiling point solvent (oleic acid) through an modified earlier reported one step method [13, 14]. In a typical experiment, iron oleate was produced by reacting 2 mmols of FeO(OH) with 8 mmols of oleic acid. The iron oleate formed in an intermediate step by the dissolution of FeO(OH) was decomposed at 320 °C using oleic acid as capping agent. The resulting particles were washed by several cycles of precipitation, centrifugation and re-dispersion in ethanol and hexane to remove the unreacted precursors and impurities. Finally, the particles were resuspended in cyclohexane and kept at 4 °C until further use.

Silica - magnetite core-shell nanoparticles

The SiO_2 coated Fe_3O_4 nanoparticles were produced in an optimized inverse microemulsion Triton-X100/hexanol/water/cyclohexane system. In a typical experiment, the microemulsion was formed by mixing the magnetic particles suspension in cyclohexane (0.89 mg Fe_3O_4 per 7 ml cyclohexane) with Triton-X100, hexanol, water, NH_4OH. After the formation of the microemulsion, TEOS was added drop wise and stirring continued for another 2 hours. The formed core-shell particles were collected by destabilizing the microemulsion with the addition of ethanol and successively washing 3-4 times with water and ethanol. The particles were finally re-suspended in water and kept at 4 °C.

DISCUSSION

The magnetic iron oxide nanoparticles with an average particle size of 9.5 (±1.2) nm were obtained by the procedure described in the Methods section. TEM images, presented in Fig. 1a, show that the particles have a narrow size distribution and are highly crystalline as observed from the HRTEM image (Fig. 1b). The average crystallite size was also calculated from XRD analysis, using Scherrer's equation and a value of 10.5 nm was obtained.

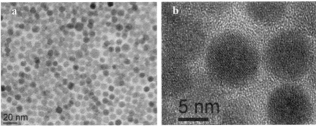

Figure 1. TEM images of the core iron oxide nanoparticles synthesised by thermal decomposition method at low magnification (a), and HRTEM (b)

A biocompatible surface is a pre-requisite for biomedical application of nanoparticles. Silica is a rather good candidate which is benign and can therefore increase the bio-compatibility of nanoparticles while preserving their intended functionality. The iron oxide nanoparticles were used as core for the fabrication of core-shell nanocomposites in an inverse (w/o) microemulsion system based on Triton-X100/hexanol/water/cyclohexane. In this method the oil phase consisted of cyclohexane and Triton-X100/hexanol is the surfactant/co-surfactant couple. The iron oxide nanoparticles are easily dispersible in cyclohexane due to the hydrophobic oleic acid capping, and this is particularly important for the microemulsion system and the subsequent preparation of magnetic core-silica shell nanocomposites. The amount of magnetic particles in microemulsion was adjusted to ensure the formation of unique core coated with silica layer with no silica nanoparticles or multicores beads formed.

In order to evaluate the influence of various experimental parameters of the synthetic process on the formation of the silica layer we investigated in this study the effect of the stirring rate and the reaction time on the growth of the silica layer. The stirring rate was varied from 400 rpm to 1200 rpm. At all stirring rates, the formation of single core core-shell nanoparticles took place (fig. 2a and fig. 2b). Even if the stirring rate has shown to play an important role of keeping the newly formed particles in the microemulsion the overall size of the silica – iron oxide nanoparticles remains constant with a slight increase in the size and polydispersity at the higher stirring rates (fig. 2c).

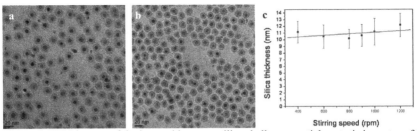

Figure 2. TEM images of the iron oxide core – silica shell nanoparticles at stirring rates of: (a) 400 rpm, (b) 1200 rpm. (c) Dependence of overall size of the core – shell nanoparticles on the stirring rate.

Especially for biomedical applications, it is essential to precisely control the thickness of the silica shell. Therefore we investigated the variation of silica shell thickness as a function of time, where the amount of TEOS and the core size were maintained constant. The formation of single core core – shell nanoparticles takes place exclusively with no empty silica particles observed at all reaction times. Also, even after a short reaction time of 2h (fig. 3a) the silica shell has a thickness of ~6 nm. The thickness of the silica shell increases linearly with time and reaches a value of ~14 nm after 17 h (fig. 3b) remaining almost constant even if the reaction time is further increased. The addition of fresh TEOS after the stabilisation of the size increased the silica thickness further which demonstrates that we can grow even further the thickness of the silica shell as shown by the size distributions in fig. 3c.

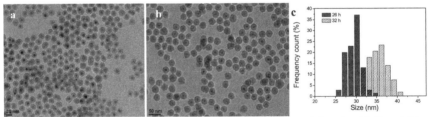

Figure 3. TEM images of the iron oxide core – silica shell nanoparticles after (a) 2h and (b) 17h reaction time. (c) Size distribution of the particles after 26h (the time of addition of fresh TEOS) and 32h reaction time.

Magnetic measurements of the SPION nanoparticles and core-shell nanocomposites by VSM analysis is represented in Fig. 4 and show that the particles are superparamagnetic (SPION) with a saturation magnetization (Ms) of 30 emu/g and no coercivity. The magnetic core size calculated from VSM (8.2 nm) is in good agreement with value obtained from TEM. Core-shell nanocomposites do show superparamagnetic character and measured Ms values are varying between 12.5, 6.5 and 2.8 emu/g for core – shell samples coated 2 h, 6h and 17 h (6 nm, 9 nm and 14 nm respectively the silica layer thickness). These nanocomposite structures are sufficiently magnetic responsive for medical imaging or targeted drug delivery applications.

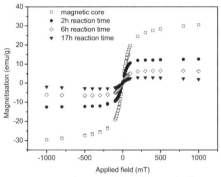

Figure 4. Magnetic measurements on the core (□) and core-shell nanocomposite spheres performed by VSM after 2h (●), 6h (◊) and 17 h (▼) of reaction.

The relaxivities of the core – shell particles with different shell thickness (6 nm, 9 nm and 14 nm respectively silica layer thickness) are summarized in Table I and are compared with two commercial T2 contrast agents: Resorvist (Schering AG, Germany) and Feridex (Advanced Magnetics, Cambridge, Massachusetts). It is to be noticed that r_2/r_1 ratios of the core shell nanoparticles obtained are ~ 3.5 fold higher than those of Resorvist at 0.47T (20 MHz) and 1.41 T (60 MHz) and ~ 6 times higher than those of Feridex at 0.47T (20 MHz) which indicates a higher T2 contrast in case our core-shell particles are used.

Table I. The relaxivities values of iron oxide core – silica shell nanoparticles obtained after different reaction times, Resorvist and Feridex reported values measured at 20 MHz (0.47 T) and 60 MHz (1.41 T) in water (37°C)

	$r_1(s^{-1}mM^{-1})$		$r_2(s^{-1}mM^{-1})$		r_2/r_1	
	20 MHz	60 MHz	20 MHz	60 MHz	20 MHz	60 MHz
6 nm coating	4.12	2.11	97.9	101.7	23.73±0.91	48.15±1.11
9 nm coating	2.18	0.79	40.47	48.58	18.55±0.46	61.42±1.56
14 nm coating	1.49	0.68	34.64	41.18	23.15±0.84	60.35±1.78
Resovist	24.9	10.9	177	190	7.10	17.4
Feridex	40	-	160	-	4	-

The cell toxicity test 3-(4,5- dimethyldiazol-2-yl)-2,5 diphenyl tetrazolium bromid (MTT) assay was performed. MTT assay is a quantitative colorimetric assay assessing the mammalian cells survival and proliferation. The cells (A549 cell line - carcinoma human alveolar basal epithelial cells) were incubated at 37 °C for 6h and 24h. The particles studied were the same series of particles on which the magnetic characterisations were performed, core – shell nanoparticles with variable silica shell thickness obtained after 2h, 6h and 17h of reaction. The particles doses used for tests were 1 μg/ml, 10 μg/ml and 100μg/ml. The preliminary cytotoxicity tests revealed that the particles were not affecting the viability and proliferation in the time interval studied. Further tests for confirming these results are undertaken and results will be reported elsewhere.

CONCLUSIONS

We presented a synthetic route of producing adjustable shell thickness of single core core – shell nanoparticles. The influence of the stirring rate and condensation time are investigated in order to finely tune the thickness of the silica shell. The magnetic nanoparticle cores exhibit a superparamagnetic behavior. The core-shell nanoparticles with different thickness of the silica shell preserved the superparamagnetic character, with an observed trend of decrease in the magnetization saturation with the increase of the silica shell thickness. The ratio r_2/r_1 for all the thicknesses of the silica shell studied is cca 3.5 times and respectively 6 times higher than the values for the commercially available T2 contrast agents investigated (Resovist and Feridex). Preliminary cell tests of the tunable core – shell particles showed no significant cytotoxicity. Thus, the presented magnetic core-silica shell nanoparticles have great potential for biomedical applications such as drug delivery systems, cell separation, and MRI contrast agents.

ACKNOWLEDGMENTS

This work is partially funded by the European Commission sixth and seventh Framework Program (INNOMED LSHB-CT-2005-518170, BIODIAGNOSTICS NMP4-CT-2005-017002, NANOMMUNE NMP-2007-SMALL-1-214281), FNRS, the FP7 (ENCITE Program) and the ARC Program 05/10-335 of the French Community of Belgium. The fellowship from Knut and Alice Wallenbergs Foundation for Dr. M. S. Toprak (No:UAW2004.0224) is thankfully acknowledged.

[1] P. Tartaj, M. del Puerto Morales, S. Veintemillas-Verdaguer et al., Journal of Physics D: Applied Physics 36, R182 (2003).

[2] Philip D. Rye, Bio/Technology 14, 155 (1996).

[3] A.-H. Lu, E. L. Salabas, and F. Schueth, Angewandte Chemie, International Edition 46, 1222 (2007).

[4] S. Laurent, D. Forge, M. Port et al., Chemical reviews 108 2064 (2008).

[5] A. K. Gupta, M. Gupta, Biomaterials 26, 3995 (2005).

[6] J. Park, K. An, Y. Hwang et al., Nature Materials 3, 891 (2004).

[7] W. Stöber, A. Fink, and E. Bohn, Journal of Colloid and Interface Science 26, 62 (1968).

[8] K. Osseo-Asare and F. J. Arriagada, Colloids and Surfaces 50 321 (1990); 69, 105 (1992); Colloids and Surfaces, A: Physicochemical and Engineering Aspects 154, 311 (1999); Journal of Colloid and Interface Science 211, 210 (1999); 218, 68 (1999).

[9] L. Yao, G. Xu, W. Dou et al., Colloids and Surfaces, A: Physicochemical and Engineering Aspects 316, 8 (2008).

[10] D. K. Yi, S. S. Lee, G. C. Papaefthymiou et al., Chemistry of Materials 18, 614 (2006).

[11] C. - W. Lu, Y. Hung, J. - K. Hsiao et al., Nano letters 7, 149 (2007).

[12] M. Zhang, B. L. Cushing, C. J. O'Connor, Nanotechnology 19, 085601/1 (2008).

[13] Yu, William W., Falkner, Joshua C. et al. Chem . Commun., 2306–2307(2004).

[14] Lin, Chun-Rong, Chiang, Ray-Kuang et al., Journal of Applied Physics 99, 08N710 (2006)

Improved Mechanical Properties of Nanocrystalline Hydroxyapatite Coating for Dental and Orthopedic Implants

Wenping Jiang[1], Jiping Cheng[2], Dinesh K. Agrawal[2], Ajay P. Malshe[1], Huinan Liu[1]
[1] NanoMech LLC, 535 W. Research Blvd, Suite 135, Fayetteville, AR 72701, USA
[2] Materials Research Institute, Penn State University, University Park, PA 16802, USA

ABSTRACT

Hydroxyapatite (HA) has been widely used as a coating material for orthopedic/dental applications due to its similar chemical composition to natural bone mineral and its capability to promote bone regeneration. It has been reported that HA with nano-scale crystalline features and controlled porosity and pore size could promote osseointegration (that is, direct bonding to natural bone). So far, a number of methods have been developed or used commercially to deposit HA on metal implants, such as electrophoretic deposition, sputter, dip coating, spin coating and plasma spray. It is, however, very challenging to produce a nanocrystalline HA coating with desirable nano-features and surface roughness as well as controlled pore size and porosity for dental/orthopedic implants. It is also necessary for nano-HA coating to have good adhesion strength to metallic substrates and sufficient mechanical properties for load-bearing conditions. Therefore, a novel hybrid coating process, combing electrostatic spray coating (ESC) technique with a novel non-conventional sintering, was developed to meet requirements for dental implants in this study. Specifically, HA nanoparticles were deposited on titanium substrates using ESC technique and the green HA coating was then sintered in a controlled condition. The produced HA coating were characterized for grain size and pore size using an environmental scanning electron microscope (ESEM), the composition and Ca/P ratio using Energy Dispersive X-ray (EDX) analysis, and crystalline phases using X-ray diffraction (XRD). The results demonstrated that a nanocrystalline HA coating with a grain size from 50 to 300 nm and a gradient of nano-to-micron pore sizes were fabricated successfully using this novel coating process. The controlled nano-scale grain size and a gradient of pore sizes are expected to promote bone cell functions and facilitate bone healing. Besides biological properties, such HA coating was also characterized for its mechanical properties, such as adhesion strength, hardness and toughness. Microscratch test results showed that the critical load of coating de-lamination reached as high as 10 N. In conclusion, this study demonstrated that it is very promising to scale up this novel hybrid coating process (ESC followed by a novel sintering process) for dental/orthopedic implant applications.

INTRODUCTION

Hydroxyapatite (HA) has been widely used as a coating material for dental/orthopedic applications due to its excellent biocompatibility and bioactivity to promote natural bone-implant integration. It has been reported that HA with nano-scale crystalline features and controlled porosity and pore size could promote osseointegration (that is, direct bonding to natural bone). So far, a number of methods have been developed or used commercially to deposit HA on metal implants, such as electrophoretic deposition, sol-gel, sputter, dip coating, spin coating and plasma spray. It is, however, very challenging to produce a high quality nanocrystalline HA coating with desirable nano-features and surface roughness as well as controlled pore size and porosity for biomedical implants. It is also necessary for nano-HA coating to have excellent adhesion strength on metallic substrates to prevent coating delamination. Therefore, in this study, a novel hybrid coating process, combining NanoMech's patented NanoSpray® coating technology with a microwave sintering process, was developed to meet mechanical and biological requirements for dental/orthopedic implants.

MATERIALS AND METHODS

HA nanoparticles were deposited on titanium (Ti) substrates using NanoMech's patented Nanospray® coating system which was developed based on the electrostatic spray coating technology (Figure 1). The HA nanoparticles were generally electrically insulating in nature and can carry the static charge over a distance of a few tens of centimeters. The HA particles were charged when they exit the powder spray gun and exposed to an electrostatic field generated by a pointed electrode, which is typically of a few tens of kilovolts. The charged particles followed the electric field lines toward the grounded objects (titanium substrates in this study) and form a uniform coating perform. The produced HA coating was then sintered in a microwave furnace for maximal consolidation without significant grain growth. The sintering of HA-coated Ti implants need to be done in an air environment in order to achieve desirable nano-HA chemistry (Ca/P ratio of 1.60 ± 0.06 to mimic natural bone mineral). The parameters for the sintering were set at a temperature of 1000-1300 °C for 5-20 minutes.

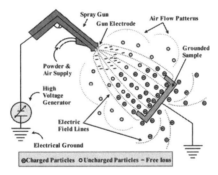

Figure 1: The schematic sketch showing the mechanism of NanoSpray® of HA nanoparticles.

The produced HA coating was characterized for grain size and pore size using an environmental scanning electron microscope (ESEM), the composition and Ca/P ratio using Energy Dispersive X-ray (EDX) analysis, and crystalline phases using X-ray diffraction (XRD). Such HA coating was further characterized for its mechanical properties, such as adhesion strength (scratch resistance), hardness and toughness. The microscratch test method is commonly used to measure the critical load and then correlate it to the coating adhesion. Thus, microscratch testing according to ASTM C1624 was carried out for the HA-coated samples. The diamond stylus was drawn on top of each sample by using an increasing load, between 10 to 120N, at constant velocity of 10 mm/min, until some well-defined failure occurred. The normal load under which the de-lamination of the coating from the Ti implants occurred was defined as critical load.

RESULTS

Figure 2 showed the results of HA nanocoatings deposited by NanoSpray® system before the microwave sintering. The HA coating surface morphology was characterized using SEM (Figure 2a-c). The chemical composition of HA coating before microwave sintering was characterized using EDX and the results are shown in Figure 2e in comparison with original HA particles (Figure 2d). The coating thickness variation was thoroughly characterized using cross sections and statistical results showed the thickness of 60 ± 2.1 μm. A representative cross sections of the deposited HA nanocoatings is shown in Figure 2f.

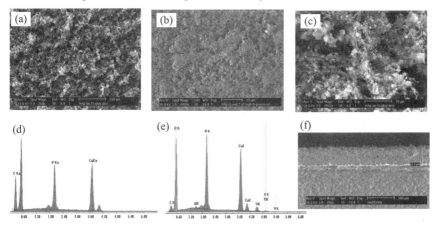

Figure 2: (a-c) SEM images of HA coating deposited by NanoSpray® system (before microwave sintering) on titanium substrates from low to high magnifications. (d) EDX results of HA particles before the coating process. (e) EDX results of HA nanocoatings on titanium substrates. (f) SEM micrograph of the cross-section of the HA nanocoating on titanium substrates to show the coating thickness (59.4 μm) and uniformity. The scale bars are 100 μm for (a) and (f), 20 μm for (b), 10 μm for (c).

After the microwave sintering, the results demonstrated that a nanocrystalline HA coating with a grain size from 50 to 300 nm and a gradient of nano-to-micron pore sizes were fabricated successfully using this novel coating process (Figure 3). The controlled nano-scale grain size and a gradient of pore size are expected to promote bone cell functions and facilitate bone healing. EDX results showed that the nano-HA coating had a Ca/P ratio of 1.6 (Figure 3d), very close to natural bone. XRD results confirmed that the nano-HA coating was highly crystalline after sintering (Figure 3e).

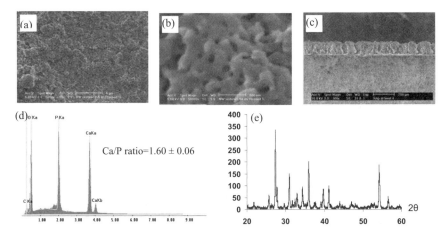

Figure 3: (a-c) SEM micrographs of HA nanocoating after microwave sintering. (a) Low magnification to show uniformity. (b) High magnification to show nano-scale features and nano-porosity. (c) Cross-section of HA coating to show the thickness of coating (~ 60 μm) and its uniformity. (d) EDX results of HA nanocoatings after microwave sintering showed the Ca/P ratio of 1.60 ± 0.06 (mimic natural bone mineral). (e) XRD results of HA nanocoatings after microwave sintering confirmed presence of highly crystalline HA.

Optical examination at the end of the microscratch test coupled with both acoustic emission response and frictional properties variation during the test provided complete insight into the coating adhesion. Microscratch test results showed that the critical load of coating de-lamination reached as high as 10 N.

CONCLUSIONS

This study demonstrated that it is very promising to scale up this novel hybrid coating process (NanoSpray® coating followed by a microwave sintering) for commercial dental/orthopedic implants. This coating process offers high deposition rate, suitability for various composite coatings, compatibility with simple and complex geometries, flexibility, low energy consumption and low cost. The results demonstrated that the application of this coating process could reduce or even eliminate the formation of amorphous phase HA, which is soluble in body fluids and results in subsequent dissolution of the material before natural bone tissue integrates. Technically, the HA nanocoatings fabricated by this coating process have the following benefits.

- Improved adhesion strength prevents delamination;
- Biomimetic chemistry to natural bone tissues (Ca/P ratio very close to natural bone);
- Large effective surface areas enhance cell attachment and growth;
- Nano-scale roughness promotes implant-tissue integration;
- Nano-to-micron pores provide more anchor sites for inducing cell activities;

- Highly crystalline HA coating reduces HA dissolution in body fluids.

ACKNOWLEDGMENTS

The authors would like to thank NIH SBIR program for financial support.